The Golden Ratio

Springer Nature More Media App

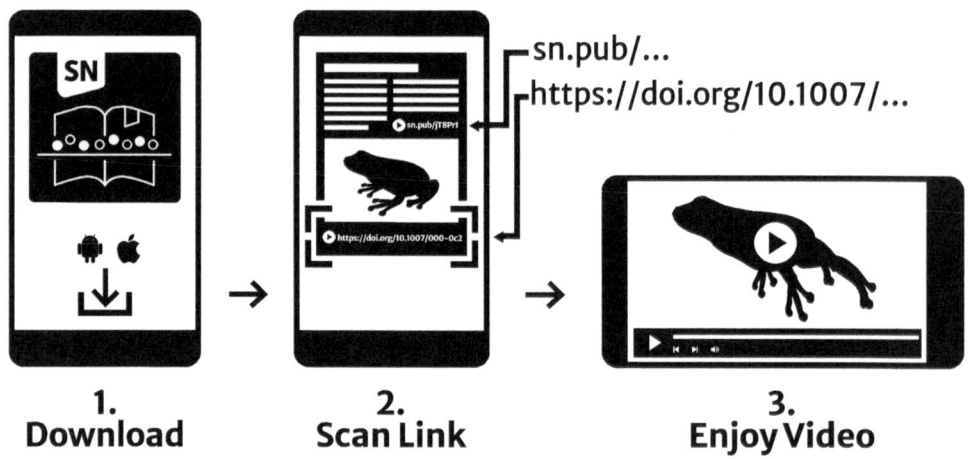

1.
Download

2.
Scan Link

sn.pub/...
https://doi.org/10.1007/...

3.
Enjoy Video

Support: customerservice@springernature.com

Hans Walser

The Golden Ratio

Geometric and Number Theoretical Considerations

 Springer

Hans Walser
Frauenfeld, Switzerland

This work contains media enhancements, which are displayed with a "play" icon. Material in the print book can be viewed on a mobile device by downloading the Springer Nature "More Media" app available in the major app stores. The media enhancements in the online version of the work can be accessed directly by authorized users.

ISBN 978-3-662-69889-1   ISBN 978-3-662-69890-7   (eBook)
https://doi.org/10.1007/978-3-662-69890-7

Translation from the German language edition: "Der Goldene Schnitt (6., bearbeitete und erweiterte Aufl.)" by Hans Walser, © Der/die Herausgeber bzw. der/die Autor(en), exklusiv lizenziert an Springer-Verlag GmbH, DE, ein Teil von Springer Nature 2013. Published by Edition am Gutenbergplatz Leipzig. All Rights Reserved.

This book is a translation of the original German edition "Der Goldene Schnitt," 7th edition, by Hans Walser, published by Springer-Verlag GmbH, DE in 2023. The translation was done with the help of an artificial intelligence machine translation tool. A subsequent human revision was done primarily in terms of content, so that the book will read stylistically differently from a conventional translation. Springer Nature works continuously to further the development of tools for the production of books and on the related technologies to support the authors.

This Springer imprint is published by the registered company Springer-Verlag GmbH, DE, part of Springer Nature.
The registered company address is: Heidelberger Platz 3, 14197 Berlin, Germany

If disposing of this product, please recycle the paper.

Preface

The Golden Ratio has been appearing since antiquity in many areas of geometry, architecture, music, art, and philosophy. The Golden Ratio is not an isolated phenomenon, but in many cases the first and thus simplest non-trivial example in the context of further generalizations.

The present book focuses on the mathematical, particularly geometric and number-theoretical aspects of the Golden Ratio. Whenever possible, a visual representation in images and geometric animations is aimed for.

In detail, the following topics are discussed: Golden Geometry, Folding and Cutting, Number Sequences, as well as Regular and Semi-Regular Bodies.

The aim is to discuss examples of the Golden Ratio on the one hand, and to point out further paths on the other. And because the Golden Ratio is often the simplest example in this chain of generalizations, it also gains a didactic significance, as the simplest non-trivial case is very often dealt with in class.

The book is aimed at students, pupils, mathematics teachers, and interested laypeople. It is modular in structure, so the individual chapters can be read independently of each other. The reader is encouraged to engage in their own geometric and algebraic activities, but also receives tips and procedural advice from the craft-creative area. Based on my experience with previous editions of this book, the topic of the Golden Ratio is often and gladly taken up in school in-depth studies such as high school graduation and maturity works or semester papers.

I owe many examples to interested readers and fellow teachers. I owe special thanks to Peter Gallin, Zurich, Johanna Heitzer, Aachen, Alfred Hoehn, Basel, Hans Rudolf Moser, Frauenfeld, Hartmut Müller-Sommer, Vechta, Jo Niemeyer, Berlin, Hartmut Rehlich, Braunschweig, Hans Rudolf Schneebeli, Wettingen, Reto Schuppli, Matzingen, Hans-Jürg Stocker, Wädenswil, Heinz Klaus Strick, Leverkusen, Bodo von Pape, Oldenburg, Anton Weininger, Landshut and Roland Wyss, Flumenthal.

The first two editions of this book were published by B.G. Teubner Verlagsge-sellschaft, Stuttgart-Leipzig, in 1993 and 1996, and the third to sixth editions were published in the Edition am Gutenbergplatz, Leipzig, in the years 2003 to 2013.

The present seventh edition at Springer Spectrum is a complete revision.

Frauenfeld   Hans Walser
October 2023

Contents

What is it about?

1

1.1   What is the "Golden Ratio"?

The *Golden Ratio* is a proportion that appears in various geometric and arithmetic figures and situations. In the figures of Fig. 1.1, which are composed of equilateral triangles, squares, regular pentagons, and circles, we see three points lying on a straight line. The point between the two outer points is not in the middle, but it always divides the distance between the outer points in the same ratio, namely in the ratio of the Golden Ratio.

But what is the Golden Ratio? Here is the definition:

A line is divided in the ratio of the Golden Ratio if the two segments relate to each other as the whole line to the longer segment.

If the whole line has the length x and the longer segment has the length 1, as shown in Fig. 1.2, this is called:

$$\frac{x}{1} = \frac{1}{x-1} \tag{1.1}$$

Thus, x is a solution of the quadratic equation $x^2 - x - 1 = 0$. This equation has the two solutions

$$x_1 = \frac{1 + \sqrt{5}}{2} \approx 1.618, \; x_2 = \frac{1 - \sqrt{5}}{2} \approx -0.618 \tag{1.2}$$

The length we are looking for x must be positive, so:

$$x = \frac{1 + \sqrt{5}}{2} \approx 1.618 \tag{1.3}$$

Supplementary Information The online version contains supplementary material available at https://doi.org/10.1007/978-3-662-69890-7_1.

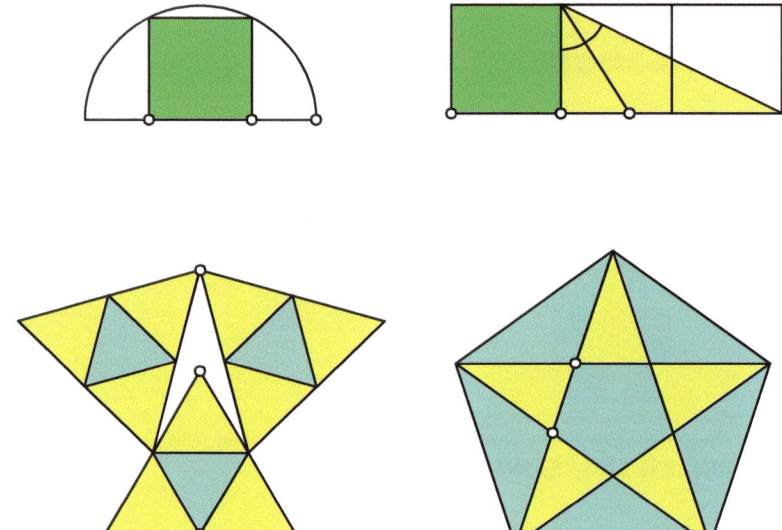

Fig. 1.1 Equal proportions

Fig. 1.2 Division in the
Golden Ratio

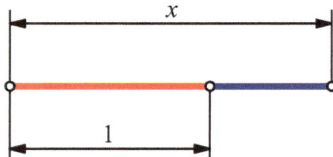

We denote this number with Φ :

$$\Phi = \frac{1 + \sqrt{5}}{2} \approx 1.618 \tag{1.4}$$

The partial ratio of the Golden Ratio has been known since antiquity as a continuous division and was discovered by the Pythagoreans on the pentagon and the dodeca-hedron (cf. [13, 25], p. 357, [34]). The Italian Franciscan monk and mathematician Luca Pacioli (ca. 1445–1517) used the term "Divina Porportione", divine ratio (cf. [13], p. 171, [26], p. 263). The term "Golden Ratio" commonly used in German today was introduced in 1835 by Martin Ohm (1792–1872). Martin Ohm was the younger brother of the experimental physicist Georg Simon Ohm (1789–1854).

The richness of aspects of the Golden Ratio means that any representation can only be exemplary. Therefore, some literature references to areas that are not discussed in detail in this book: The architectural and artistic aspect is in the foreground in [6, 8, 9, 11, 17, 28] and [35] are dedicated to the aesthetic side of mathematics. [2] and [33] provide clear introductions to various areas of the Golden Ratio. Finally, [13, 20] and [21] provide a detailed historical overview of the history of the Golden Ratio and its occurrence in literature.

1.2 Designations

According to Sect. 1.1 is:

$$\Phi = \frac{1 + \sqrt{5}}{2} \approx 1.618 \tag{1.5}$$

For the reciprocal $\frac{1}{\Phi}$ we get:

$$\frac{1}{\Phi} = \frac{2}{1 + \sqrt{5}} = \frac{-1 + \sqrt{5}}{2} \approx 0.618 \tag{1.6}$$

The symbol Φ was proposed around 1909 by the American mathematician Mark Barr as a reference to the Greek sculptor Phidias (around 480 BC to 430 BC), whose name in the Greek alphabet begins with Φ. In the literature, other designations are used instead of Φ, in particular (as in earlier editions of this book) $\tau = \Phi$ and $\rho = \frac{1}{\Phi}$.

For the two numbers Φ and $\frac{1}{\Phi}$ the following relationships apply:

$$\Phi + \frac{1}{\Phi} = \sqrt{5}, \Phi - \frac{1}{\Phi} = 1, \Phi^2 - \Phi = 1, \left(\frac{1}{\Phi}\right)^2 + \frac{1}{\Phi} = 1 \tag{1.7}$$

The quadratic equation $x^2 - x - 1 = 0$ has the two solutions $x_1 = \Phi$ and $x_2 = -\frac{1}{\Phi}$.

The quadratic equation $x^2 + x - 1 = 0$ has the two solutions $x_1 = \frac{1}{\Phi}$ and $x_2 = -\Phi$.

We will encounter these two quadratic equations repeatedly in the following; they are the two key equations for the golden ratio.

In a division in the ratio of the golden section, we have a long and a short part. The longer part is often referred to as Major, the shorter as Minor see Fig. 1.3. The terms Major and Minor are related to each other, just like mother and daughter.

Example: We draw an equilateral triangle and a regular pentagram into the same circle Fig. 1.4. The inradius of the two figures are in the ratio of Major and Minor. I owe this figure to Anton Weininger, Landshut.

The proportion of the Minor to the Major is $\frac{1}{\Phi} \approx 0.618 = 61.8\%$.

Major   Minor

Fig. 1.3 Longer and shorter part

Fig. 1.4 Triangle and Pentagram

1.3   Continuous Division

The Fig. 1.5 illustrates the principle of continuous division. We start with a red circle, whose radius we consider as the major. Into this circle, we draw a second, blue circle with a left stop, whose radius is the associated minor. The radius of the blue circle is therefore about 61.8% of the radius of the red circle.

And now we change the colors. We now color the red circle yellow and the blue one red. This means that we now consider its radius as the major. To this, we now draw a third, even smaller and again blue circle, whose radius is the associated minor. We thus copy the radius ratios of the starting figure.

The trick of continuous division is now that we can push this smallest circle to the right, and it fits exactly into the yellow moon.

And this only works with the Golden Ratio. If we reduce the radii with 60% instead of the Golden Ratio $\frac{1}{\Phi} \approx 0.618 = 61.8\%$, the last circle can no longer be fitted exactly Fig. 1.6. It is too small.

With a reduction to 65%, on the other hand, the last circle becomes too large Fig. 1.7.

The Fig. 1.8A shows a play around continuous division.

Fig. 1.5 Continuous Division

Fig. 1.6 Too small

Fig. 1.7 Too large

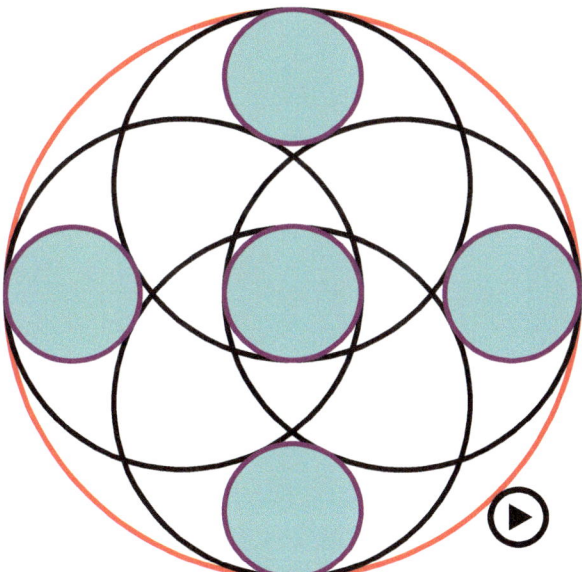

Fig. 1.8 Continuous Division (▶ https://doi.org/10.1007/000-caq)

Golden Geometry

2

2.1   Constructions of the Golden Ratio

There are many constructions of the Golden Ratio. Here is a small selection.

2.1.1   The Classic Construction

The Fig. 2.1 shows the classic construction of the Golden Ratio: In a right-angled triangle ABC with the catheti $a = 1$ and $b = \frac{1}{2}$ a circle with the center A and the radius $b = \frac{1}{2}$ is drawn; this intersects the straight line AB at the inner intersection point D and at the outer intersection point E.

With the Pythagorean theorem, we get:

$$|BD| = \sqrt{\frac{5}{4}} - \frac{1}{2} = \frac{\sqrt{5} - 1}{2} = \frac{1}{\Phi} \tag{2.1}$$

$$|BE| = \sqrt{\frac{5}{4}} + \frac{1}{2} = \frac{\sqrt{5} + 1}{2} = \Phi \tag{2.2}$$

The Fig. 2.2 shows three examples of how the terms Major and Minor can be incorporated into the figure.

We now replace in Fig. 2.1 the cathetus $b = \frac{1}{2}$ with a cathetus of length $b = \frac{n}{2}$ with $n \in \mathbb{N}$ (Fig. 2.3 for $n = 3$). Then the difference is $|BE| - |BD| = n$, since the circle around A has the diameter n. The point B has the power 1 with respect to this circle, thus $|BD|$ and $|BE|$ are reciprocals of each other. For each $n \in \mathbb{N}$ we thus

Supplementary Information The online version contains supplementary material available at https://doi.org/10.1007/978-3-662-69890-7_2.

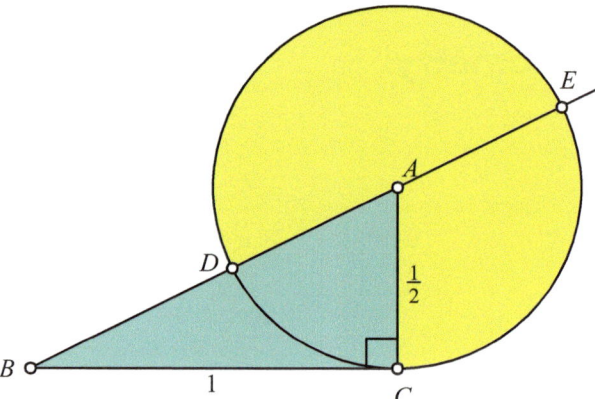

Fig. 2.1 Construction of the Golden Ratio

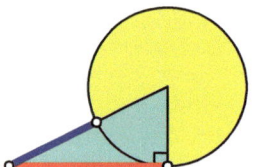 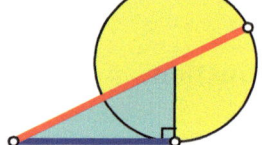 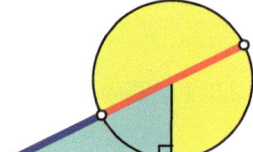

Fig. 2.2 Major and Minor

Fig. 2.3 Generalization

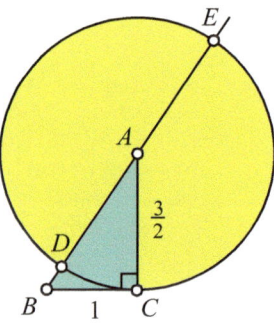

obtain two lengths, which are reciprocals of each other and differ by the natural number n, that is, they have the same decimal places. It is:

$$|BD| = \frac{-n + \sqrt{4 + n^2}}{2} \tag{2.3}$$

$$|BE| = \frac{n + \sqrt{4 + n^2}}{2} \tag{2.4}$$

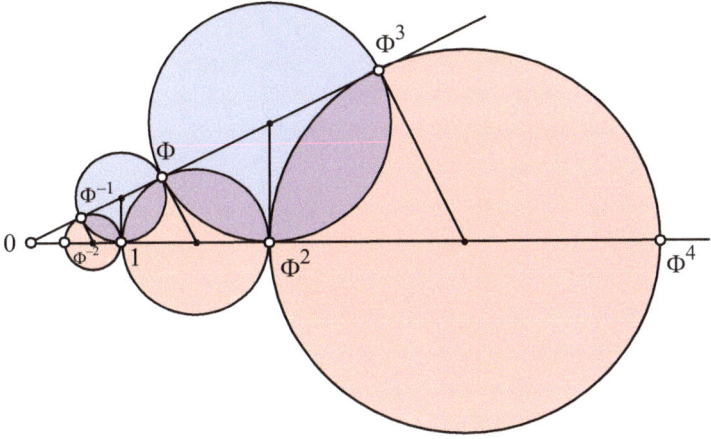

Fig. 2.4 Powers

For $n = 3$ we obtain, for example:

$$|BD| = \frac{-3 + \sqrt{13}}{2} \approx 0.302775637731995 \tag{2.5}$$

$$|BE| = \frac{3 + \sqrt{13}}{2} \approx 3.302775637731995 \tag{2.6}$$

The repetition of the construction of Fig. 2.1 leads to a geometric sequence with the quotient Φ (Fig. 2.4).

2.1.2   Construction with Bisectors

Figure 2.5 shows a construction method for the Golden Ratio with bisectors. We draw in a right-angled triangle ABC with the catheti $a = 2$ and $b = 1$ the inner and the outer bisector of the angle α. These intersect the carrier line of the cathetus a at the inner intersection point A_{-1} and at the outer intersection point A_{+1}.

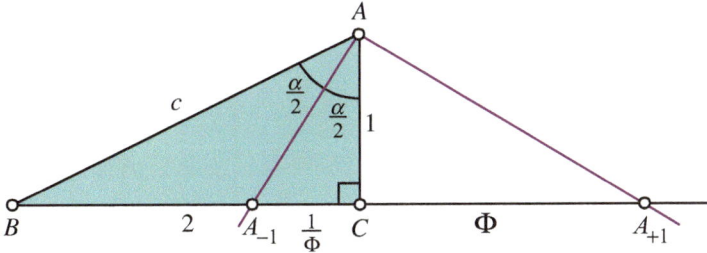

Fig. 2.5 Construction with Bisectors

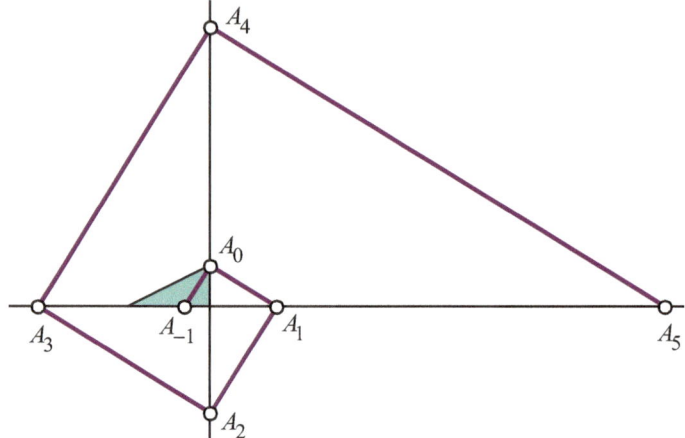

Fig. 2.6 Angular Golden Spiral

Since the internal bisector of the angle α divides the side a in the ratio of the sides b and c, $|CA_{-1}| : |BA_{-1}| = 1 : \sqrt{5}$ is obtained. Together with $|CA_{-1}| + |BA_{-1}| = 2$, this results in $|CA_{-1}| = \frac{1}{\Phi}$. Analogously, or also using the altitude theorem, we calculate the length $|CA_{+1}| = \Phi$.

The iteration of the construction of Fig. 2.5 leads to the angular Golden Spiral of the Fig. 2.6. This is a so-called logarithmic Spiral, as the distances from the center grow exponentially (cf. [43]).

Example 1 An isosceles Triangle is inscribed in a square with side length 2 (Fig. 2.7). How large is its inradius?

By drawing in the bisector, we recognize the situation of Fig. 2.5. The inradius is $\frac{1}{\Phi}$. We can now also fit in a rectangle with the aspect ratio of the Golden Section, a Golden Rectangle, that is.

Example 2 In the Pythagorean triangle with the leg lengths 3 and 4 and the hypotenuse length 5, we draw the incircle (Fig. 2.8). This has a radius of 1. On the angle bisector originating from the lower left corner, we find the ratio of the

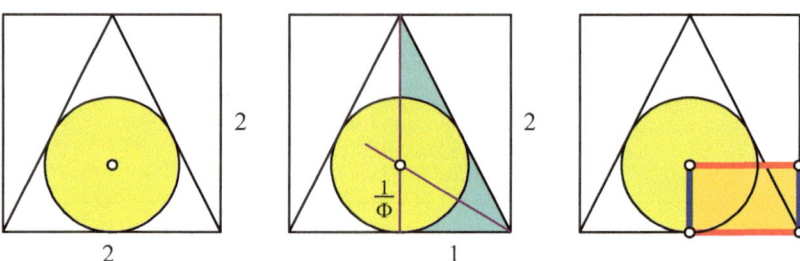

Fig. 2.7 Inradius

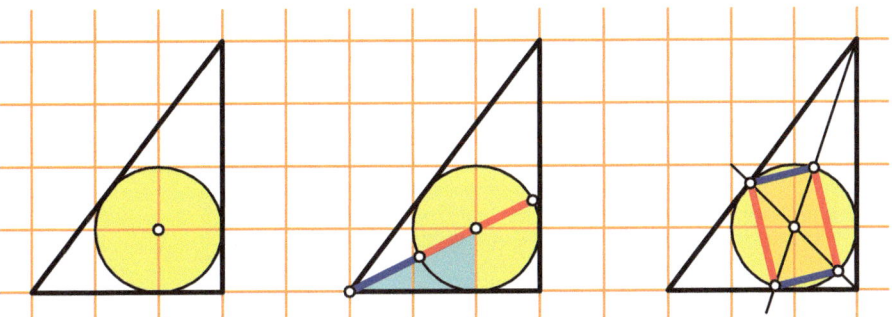

Fig. 2.8 Pythagorean triangle

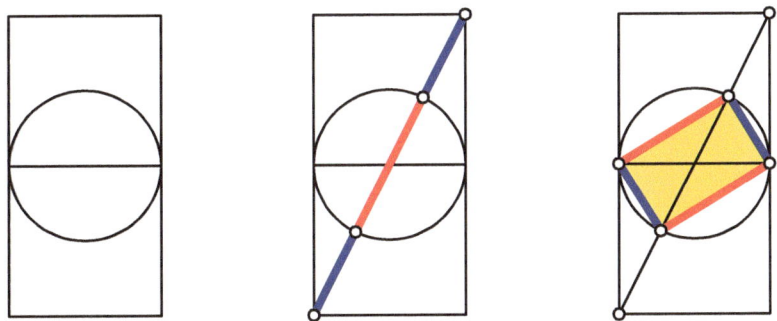

Fig. 2.9 Two squares and a circle

Golden Section. The background is again the construction of Fig. 2.1. The other two angle bisectors intersect the incircle at four points, which form a Golden Rectangle ([15], p. 87).

2.1.3   Circle and Squares

Constructions of the Golden Section and the Golden Rectangle are presented, which work with squares and a circle. In Figs. 2.9 and 2.10, two squares and a circle are used each time.

In Fig. 2.11, three squares are needed, in Fig. 2.12, five squares.

We can replace the squares of Fig. 2.12 with parallelograms (Fig. 2.13). The squares are distorted. Orthogonal square sides are distorted differently. Instead of a circumcircle of the figure, a circumellipse results.

On the diagonal in the middle, there is still a subdivision in the order Minor-Major-Minor. Distances, which lie on the same straight line or are parallel, are distorted equally, so that their length ratio is preserved. However, the Golden Rectangle of Fig. 2.12 is distorted into a parallelogram, whose side lengths are no longer in the ratio of the Golden Section.

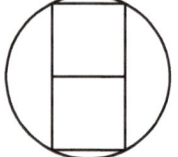 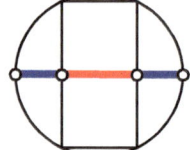 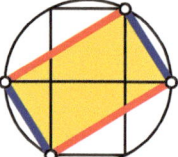

Fig. 2.10 Two squares and a circle

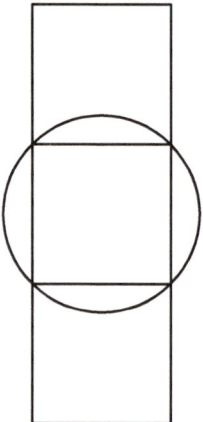 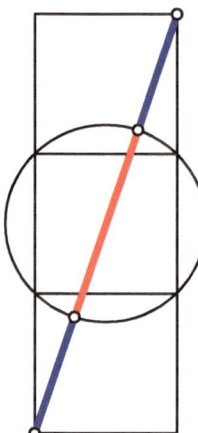 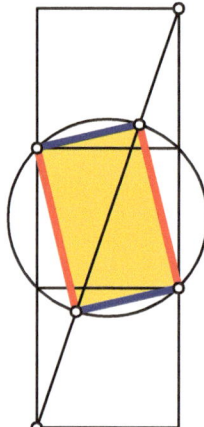

Fig. 2.11 Three squares and a circle

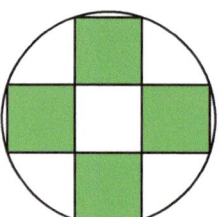 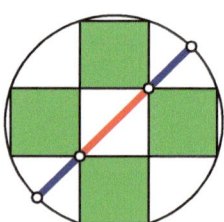 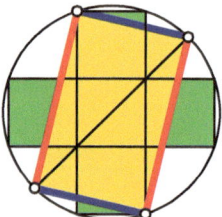

Fig. 2.12 Five squares and a circle

 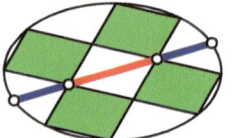 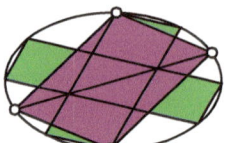

Fig. 2.13 Parallelograms instead of squares

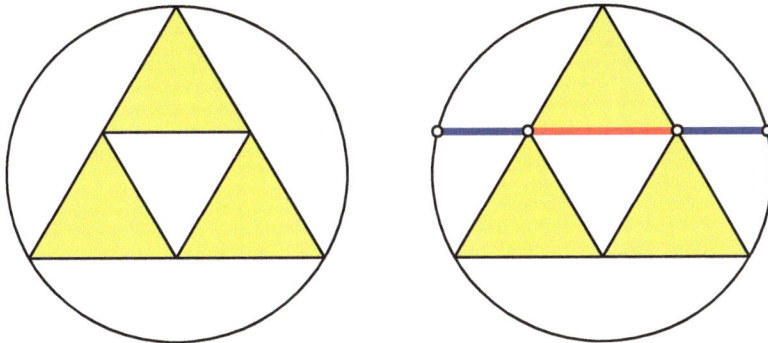

Fig. 2.14 Odom's Construction

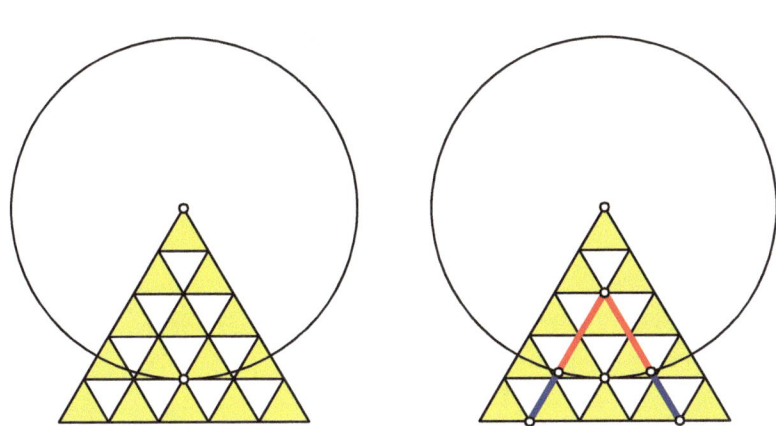

Fig. 2.15 Ferris wheel

2.1.4 Circle and Triangles

Constructions with equilateral triangles and a circle.

The construction according to Fig. 2.14 goes back to George Phillips Odom (1941–2010).

The Ferris wheel construction (Fig. 2.15) also works with equilateral triangles.

We can assemble congruent irregular triangles into a skewed star (Fig. 2.16) and provide it with a circumellipse. This also results in a construction possibility for the golden ratio. It is a variant of Odom's construction.

2.1.5 Circle, Squares and Triangles

We combine squares and triangles and add a circle. In Fig. 2.17 there are two squares and three triangles, in Fig. 2.18 two squares and four triangles.

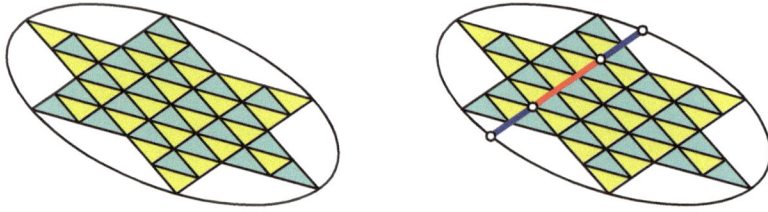

Fig. 2.16   Skewed Star

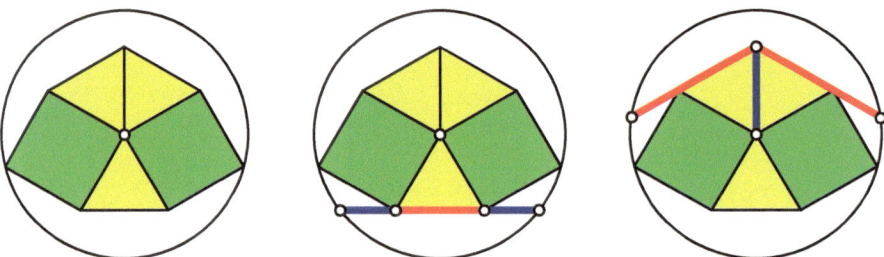

Fig. 2.17   Two Squares and Three Triangles

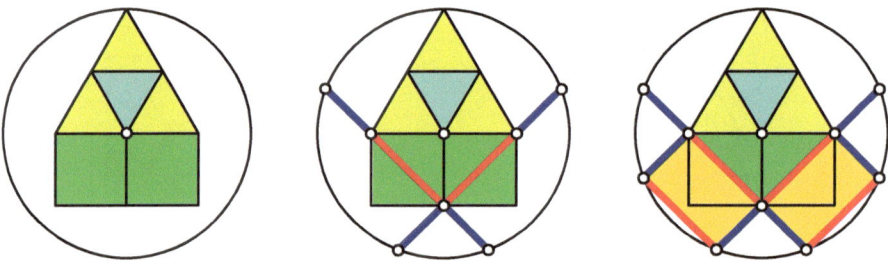

Fig. 2.18   Two Squares and Four Triangles

2.1.6   A Construction with the Compass Alone

Theconstruction of Fig. 2.19 works with five circular arcs (cf. [16]). The three white points are in the ratio of the golden section.

For the proof, however, we must think of straight lines (Fig. 2.20).

We set the radius of the smallest circles to 1. Thus, $\overline{PB} = 1$ and $\overline{PC} = 2$ as well as $\overline{MB} = \frac{1}{2}\sqrt{3}$ and $\overline{MC} = \frac{1}{2}\sqrt{15}$. This results in:

$$\frac{\overline{AC}}{\overline{AB}} = \frac{\frac{1}{2}\sqrt{3} + \frac{1}{2}\sqrt{15}}{\sqrt{3}} = \frac{1 + \sqrt{5}}{2} = \Phi \tag{2.7}$$

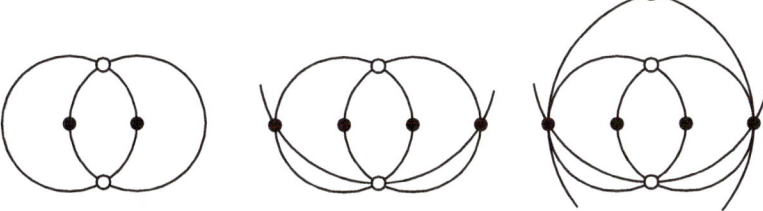

Fig. 2.19 Construction with five circles

Fig. 2.20 Proof figure

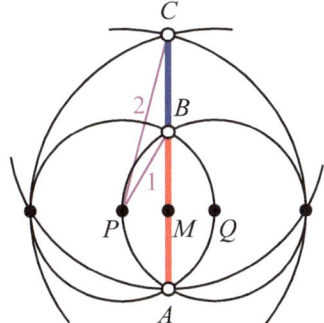

From the proof figure (Fig. 2.20) and the associated calculations, it follows that in the figures of Fig. 2.21 the three white points stacked vertically on top of each other are in the ratio of the golden section. Additionally, Major and Minor are drawn in.

2.1.7   A Construction in the Right-Angled Triangle

In any right-angled trianglewe draw the midpoints of the two catheti and transfer the height perpendicular to the hypotenuse onto the catheti (Fig. 2.22). We then draw a circle around the midpoint of one cathetus through the point transferred with the height on the other cathetus. We proceed in the same way with the second cathetus.

The intersection points of the two circles lie on the extended triangle height and divide it in the ratio of the golden section. The triangle height is the Major, the attached sections are Minors.

Since the construction applies to any right-angled triangle, we can vary the corner with the right angle on the Thales circle (Fig. 2.23).

The proof is done computationally. We work with the designations of Fig. 2.24. We set the height to 1. The hypotenuse section p from the height base point H to the corner A is variable.

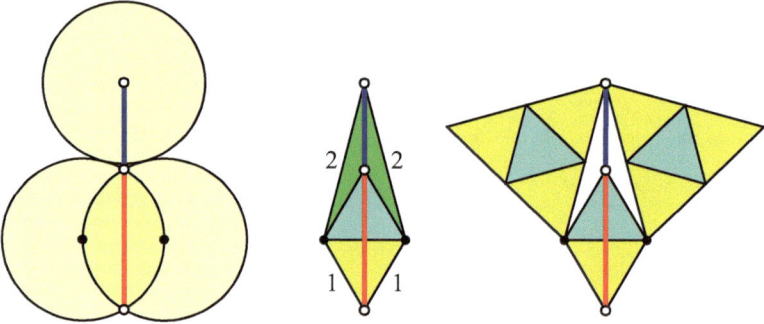

Fig. 2.21 Ratio of the Golden Section

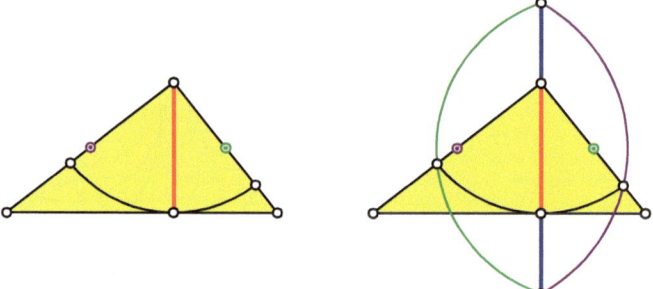

Fig. 2.22 In the right-angled triangle

The point D as the center of the line AC has the coordinates $\left(-\frac{p}{2}, \frac{1}{2}\right)$. For the section CD we get:

$$\overline{CD}^2 = \frac{1}{4}\left(p^2 + 1\right) \tag{2.8}$$

The triangle CDE is right-angled and has the cathetus $\overline{CE} = 1$ according to the construction. Thus, we get for r:

$$r^2 = \overline{CD}^2 + 1 = \frac{1}{4}p^2 + \frac{5}{4} \tag{2.9}$$

We intersect the altitude line with the circle around D through E. The intersection point F has the coordinates $(0, -f)$, where f is still unknown.

The triangle DFG is also right-angled and has the catheti $\frac{1}{2}p$ and $\frac{1}{2} + f$. This results in r:

$$r^2 = \frac{1}{4}p^2 + \left(\frac{1}{2} + f\right)^2 \tag{2.10}$$

Fig. 2.23 Variation of the right-
angled triangle
(▶ https://doi.org/10.1007/000-cas)

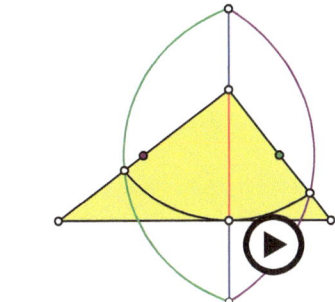

Fig. 2.24 Designations

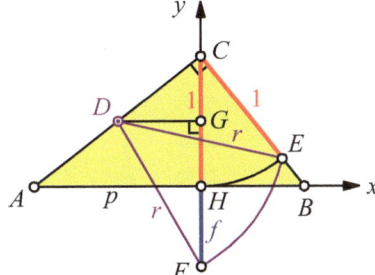

Equating the two pieces of information about r yields:

$$\frac{1}{4}p^2 + \frac{5}{4} = \frac{1}{4}p^2 + \left(\frac{1}{2} + f\right)^2 \tag{2.11}$$

We see that the variable size p "drops out". What remains is the quadratic equation:

$$f^2 + f - 1 = 0 \tag{2.12}$$

This equation has the solutions $-\Phi$ and $\frac{1}{\Phi}$.

If we now carry out the same considerations for the other cathetus of the original right-angled triangle, we arrive at the same point F. The lower intersection of the two circles according to the construction (Fig. 2.22) is therefore actually on the line through the height and provides the minor opposite the height.

For reasons of symmetry (parallel to the hypotenuse at half the triangle height as the axis of symmetry), this also applies to the upper intersection point.

In the special case of an isosceles right-angled triangle, the figure of Fig. 2.25 results. We can also fit a square into it.

In this special case, we can also work with equilateral triangles.

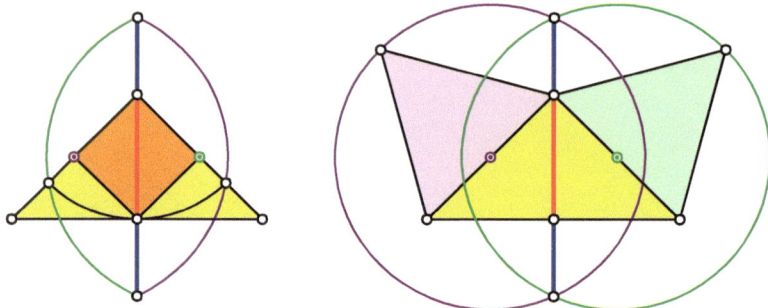

Fig. 2.25 Special case. Equilateral triangles

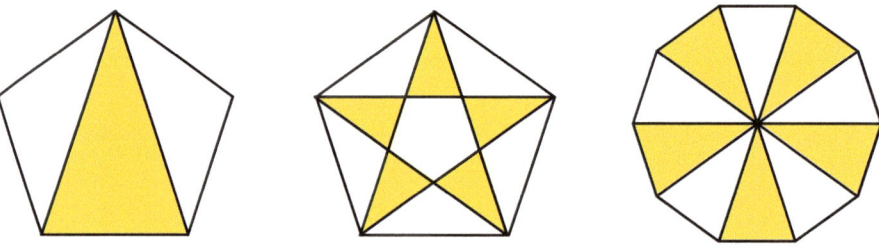

Fig. 2.26 Sharp Golden Triangle

2.2   The Regular Pentagon and the Regular Decagon

In the regular pentagon (Fig. 2.26) and in related figures such as the so-called pentagram formed from the diagonals of the regular pentagon or the regular decagon, the Golden Ratio appears in many places. A key figure is the isosceles triangle with the apex angle $36°$, the so-called *sharp Golden Triangle*.

This sharp Golden Triangle has the base angles $72°$ (Fig. 2.27); the bisector of a base angle therefore separates a similar triangle *DAB* from the whole triangle.

The remaining triangle *BCD*, the so-called *bluntGolden Triangle*, is also isosceles. We can fit a closed equilateral zigzag line with five sections into the sharp Golden Triangle.

By normalizing the base length $c = 1$ of the sharp golden triangle *ABC*, the leg length a results from the similarity of the triangles *ABC* and *DAB* the condition:

$$\frac{a}{1} = \frac{1}{a - 1} \tag{2.13}$$

This results in the quadratic equation $a^2 - a - 1 = 0$ with the positive solution $a = \Phi$ (Fig. 2.28).

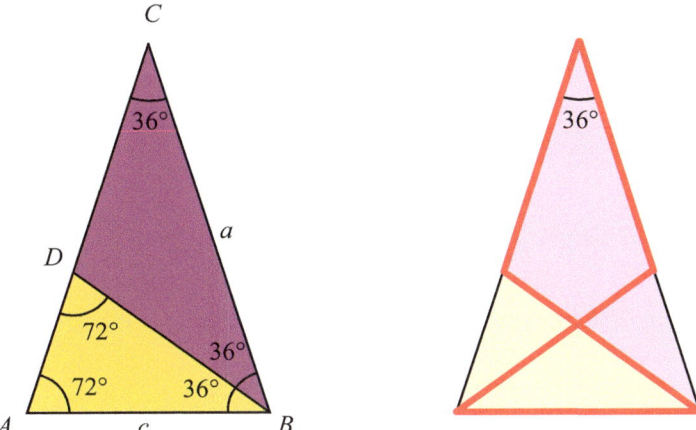

Fig. 2.27 Division of the sharp Golden Triangle. Zigzag path

Fig. 2.28 Side ratios

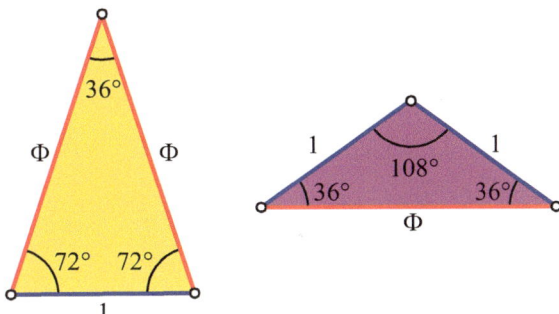

If we, conversely, normalize the leg length in the sharp golden triangle to $a = 1$, we obtain the base length $c = \frac{1}{\Phi}$. In the obtuse golden triangle with the base angle of 36°, the side lengths are in reversed order. Thus, in the regular pentagon, the sides and the diagonals are in the ratio of the golden section.

In the regular decagon (Fig. 2.29) with the circumradius 1, the side AB has the length $\frac{1}{\Phi}$ and the diagonal AD has the length Φ.

In Fig. 2.30, three different construction methods for the regular pentagon are indicated. The first two methods work according to Fig. 2.1, the third according to Fig. 2.5.

2.2.1 Approximate constructions for the regular pentagon

For a given line segment AB, the regular pentagon with this line segment as a side is sought. Below are two historical approximate constructions.

Fig. 2.29 Regular Decagon

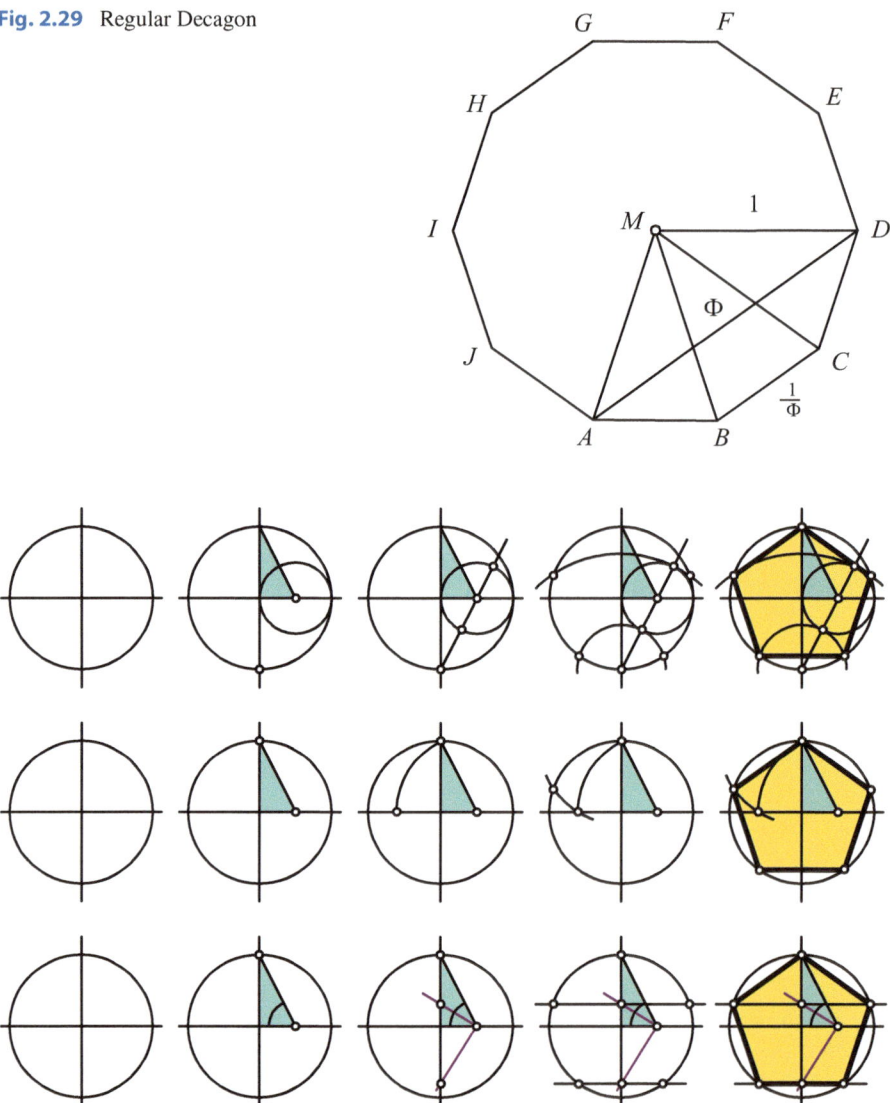

Fig. 2.30 Three construction methods for the regular pentagon

Fig. 2.31 describes a method that is said to go back to Leonardo da Vinci (1452–1519). The pentagon is neither equilateral nor equiangular. The deviations from the regular pentagon are up to 1,8 %.

The sequence of figures 2.32 shows a method, attributed to Albrecht Dürer (1471–1528). The pentagon is equilateral, but not equiangular. The angle at the corner A is about 0,3 % too large. The pentagon has no circumcircle.

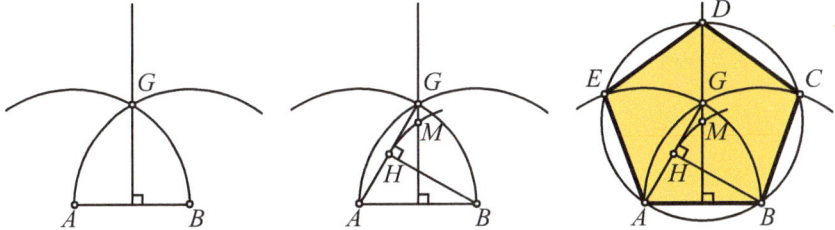

Fig. 2.31 Approximate construction by Leonardo da Vinci

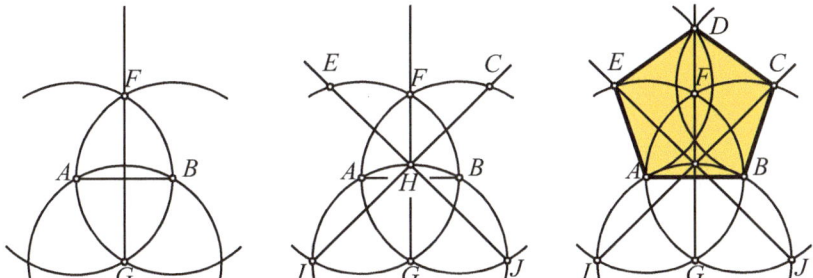

Fig. 2.32 Approximate construction by Albrecht Dürer

2.2.2   Grid Geometry

Almost every graphic software has an option where only points in a square grid (square grid, grid) can be drawn. Mathematically, this means that only points with integer coordinates are allowed. Unfortunately, it is impossible to draw the five vertices of a regular pentagon in the square grid.

We prove this indirectly by assuming that such a pentagon exists, and showing that this assumption leads to a contradiction. For this, we need the following fact: For two grid points P and Q there is another grid point R such that the triangle PQR is right-angled isosceles with the right angle at Q (Fig. 2.33).

For our indirect proof, we now assume that we have a regular pentagon $A_0A_1A_2A_3A_4$, whose corners are all grid points. We now extend each pentagon side $A_{i-1}A_i$ to a right-angled isosceles triangle $A_{i-1}A_iB_i$ with the right angle at A_i (Fig. 2.34). As a result, we obtain a regular pentagon $B_0B_1B_2B_3B_4$, which is smaller and entirely within the pentagon $A_0A_1A_2A_3A_4$, and whose corners are also grid points.

So, we can find a smaller regular pentagon for every regular pentagon in grid geometry, whose corners are also grid points. By iterating this process, we obtain a sequence of ever-shrinking pentagons, whose side lengths form a geometrically decreasing null sequence. Eventually, there is such a small pentagon that it falls through the meshes of the square grid. This contradicts the fact that its vertices would have to be grid points.

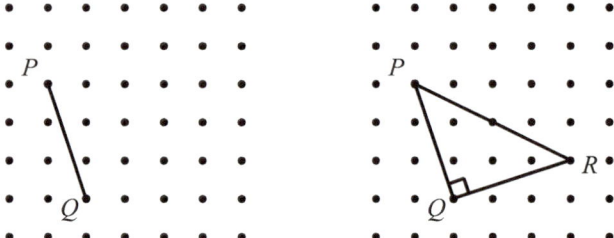

Fig. 2.33 Installation of a right angle

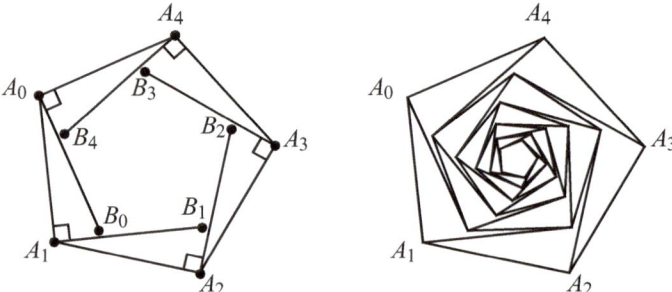

Fig. 2.34 Smaller regular pentagons

So, there is no regular pentagon in the square grid.

Similarly, it can be shown that there is no regular polygon with a number of vertices $n \geq 5$ in the square grid.

This proof method is referred to as reductio ad absurdum (reduction to a contradiction).

Of course, there are integer approximation solutions: The five points with the integer coordinates $(0,20)$, $(\pm 19,6)$, $= (\pm 12, -16)$ approximate a regular pentagon with the origin as the center.

2.3   The Golden Rectangle

We understand the Golden Rectangle to be the rectangle with the aspect ratio of the Golden Section (Fig. 2.35).

2.3.1   Unit Square and Golden Rectangle

We divide a square with side length 1 into four congruent right-angled triangles according to Fig. 2.36 and reassemble these triangles.

Fig. 2.35 Golden Rectangle

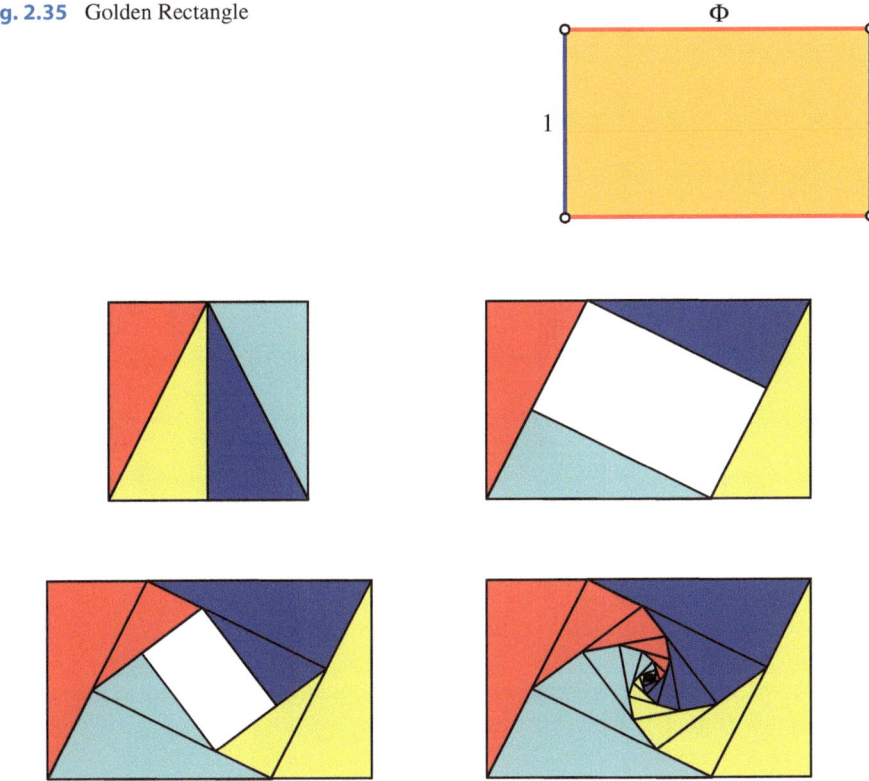

Fig. 2.36 Disassemble and reassemble. Spirals

Then a larger rectangle is created, but it has a rectangular hole. Both are Golden Rectangles, the large outline rectangle has the length Φ and the width 1, the hole rectangle has the length 1 and the width $\frac{1}{\Phi}$ (cf. [32], p. 90).

We can now fill the hole rectangle with a reduced and rotated copy of the large outline rectangle. The reduction factor is $\frac{1}{\Phi}$, the rotation angle is the small acute angle of the right-angled triangles, so the angle $\arctan\left(\frac{1}{2}\right) \approx 26{,}57°$. The rotation is clockwise. The iteration of this process yields a figure with four spirals. Two of the spirals are congruent, namely red/yellow and blue/light blue (cf. [42]).

2.3.2 Constructions of the Golden Rectangle

Starting from a square, the Golden Rectangle can be constructed very simply (Fig. 2.37). The center of the arc is the midpoint of the lower square side.

In a grid we can draw the corners of a Golden Rectangle with a single compass stroke (Fig. 2.38).

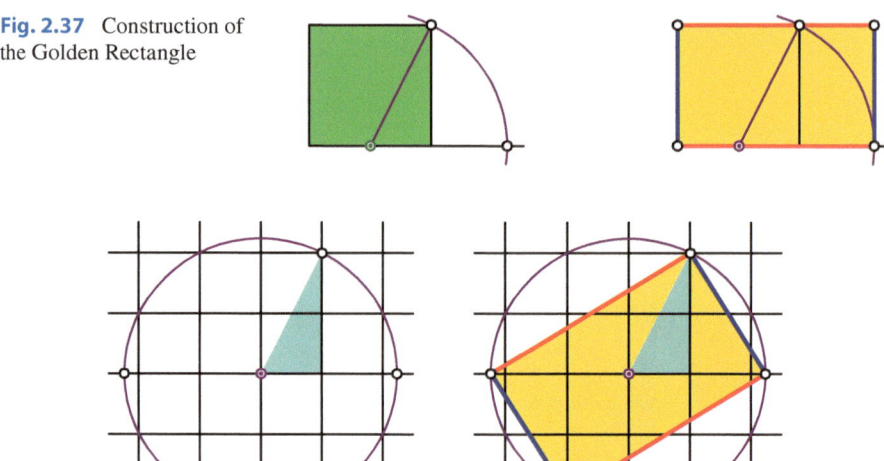

Fig. 2.37 Construction of the Golden Rectangle

Fig. 2.38 In the grid

In the figure of Fig. 2.39 we find at various places a division in the ratio of the Golden Section.

The red zigzag line oriented along the diagonal of the Golden Rectangle consists of three equal lengths (Fig. 2.40).

2.3.3 Divisions of the Golden Rectangle

First, we are looking for a rectangle with the following property: After cutting off a square from the rectangle, a remaining rectangle should be left, which is similar to the original rectangle (Fig. 2.41). The original rectangle has the length x and the width 1. If we cut off a square with the side length 1, a remaining rectangle with the length 1 and the width $x - 1$ remains. From the similarity of the original rectangle to the remaining rectangle, we get

$$\frac{x}{1} = \frac{1}{x - 1} \qquad (2.14)$$

and thus the quadratic equation:

$$x^2 - x - 1 = 0 \qquad (2.15)$$

This equation has the positive solution $x = \Phi$. The rectangle we are looking for is therefore the Golden Rectangle.

Since the remaining rectangle is again a Golden Rectangle, another square can be cut off, so that the second order remaining rectangle is again a Golden Rectangle. The iteration of this cutting process yields a sequence of squares, which

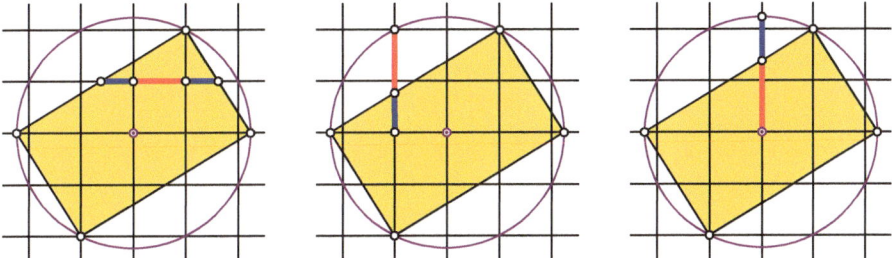

Fig. 2.39 Proportional division of the Golden Section

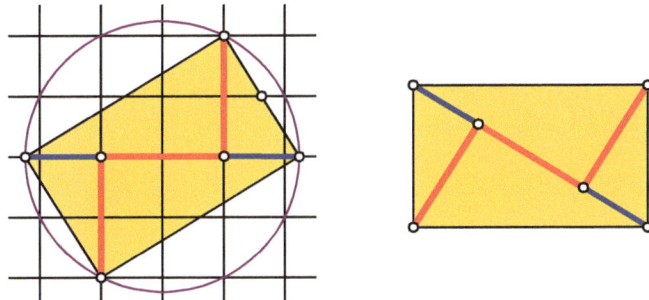

Fig. 2.40 Equilateral zigzag line

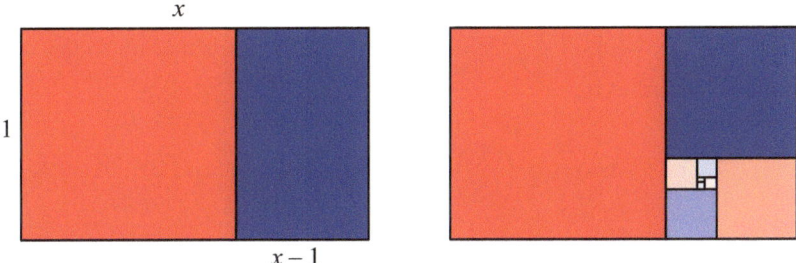

Fig. 2.41 Division of the Golden Rectangle

exhaust the original rectangle and whose sides form a geometric sequence with the quotient $\frac{1}{\Phi}$. In Fig. 2.41, these squares are arranged in a spiral.

After drawing a diagonal in the Golden Rectangle and in the first remaining rectangle, the division of the Golden Rectangle according to Fig. 2.42 can be done very easily (see [17], p. 67). The intersection of the two diagonals is at the center of the spiral.

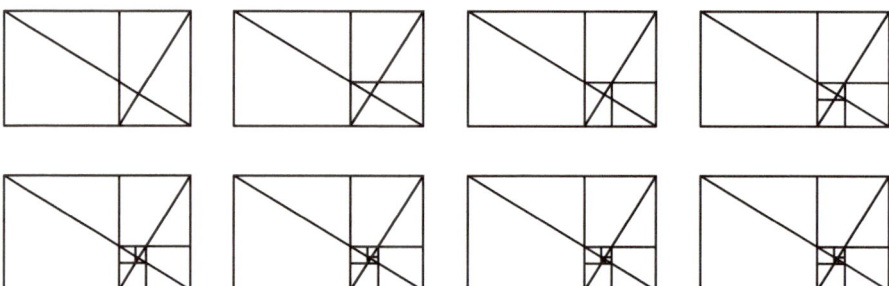

Fig. 2.42 Division process using the diagonals

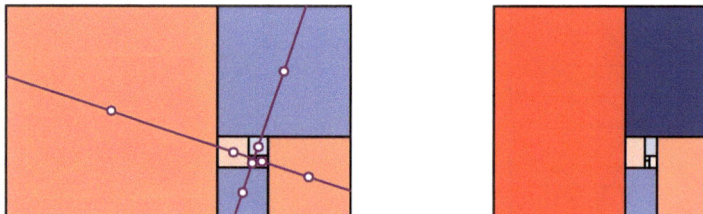

Fig. 2.43 Centers. Exhaustion of a square

The centers of the squares lie on two orthogonal lines with the slopes 3 and $-\frac{1}{3}$ (Fig. 2.43). The intersection of the two lines is the center of the spiral.

For a golden rectangle with length Φ and width 1, the side lengths of the squares are $1, \Phi, \left(\frac{1}{\Phi}\right)^2, \left(\frac{1}{\Phi}\right)^3, \ldots$. Since the square areas add up to the rectangle area, the relationship is:

$$\Phi = 1 + \left(\frac{1}{\Phi}\right)^2 + \left(\frac{1}{\Phi}\right)^4 + \left(\frac{1}{\Phi}\right)^6 + \cdots \tag{2.16}$$

Fig. 2.41 shows an exhaustion of the golden rectangle by squares. Conversely, we can also exhaust a square by golden rectangles (Fig. 2.43). This is an affine distortion of Fig. 2.41.

Fig. 2.44 shows a subdivision of the golden rectangle by smaller squares and golden rectangles. The subdivision continues infinitely to the right.

If we count the golden rectangles column by column, we get the numbers: 1, 1, 2, 3, 5, 8, 13, …. These are the Fibonacci numbers.

The column sums of the numbers of the blue squares are: 0, 1, 1, 2, 3, 5, 8, …, also the Fibonacci numbers.

In the columns from left to right, the following recursions occur:

Fig. 2.44 Subdivision of the golden rectangle

(I) A golden rectangle has a golden rectangle at the top right and a blue square at the bottom right as a neighbor.

(II) A blue square has a golden rectangle as a neighbor on the right.

Figure 2.45 shows a kinematic approach to the golden rectangle.

2.3.4 Spirals in the Golden Rectangle

Figure 2.46 shows two spirals fitted into the subdivided golden rectangle spirals, which look almost identical.

The left spiral is a logarithmic spiral ([43], p. 13 f.) through the square corners. It is sometimes referred to as the Golden Spiral ([1], p. 55, [5], p. 204, [23]). This spiral is said to be attributed to Johannes Kepler (1571–1630).

The right spiral consists of quarter circles inscribed in the squares quarter circles. The radii of the quarter circles decrease exponentially towards the inside with the factor $\frac{1}{\Phi}$. In an extended sense, this can also be referred to as a logarithmic spiral.

While the quarter circles each run within the corresponding square, the logarithmic spiral (the Golden Spiral thus) each goes a little (barely noticeable to the eye) beyond the edge of the square. In the Golden Spiral, the curvature increases continuously from the outside to the inside. With the quarter circles, the radius and thus the curvature change abruptly at the transition points of the square corners. This discontinuous curvature behavior is particularly visible at the transition point at the upper rectangle edge.

Fig. 2.45 Kinematic approach
(▶ https://doi.org/10.1007/000-car)

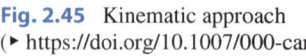

Fig. 2.46 Spirals in the Golden Rectangle

Figure 2.47 shows on the left the overlay of the two curves. The differences are minimal, but noticeable. On the right, we see a colored version of the version with quarter circles.

The fractal of Fig. 2.48 is made up of spirals with quarter circles.

If we replace the quarter circles with their complementary arcs(three-quarter circles), we get the "large spiral" on the right in Fig. 2.49. Admittedly, compared to the corresponding real logarithmic spiral on the left, it does not look particularly elegant.

Figure 2.50 illustrates the situation with three-quarter circles.

By drawing a diagonal in each square of the subdivision of the Golden Rectangle, an angular spiral results (Fig. 2.51).

2.3.5   Existence of irrational numbers

The Euclidean algorithm
The greatest common divisor of two natural numbers a and b can be calculated using the Euclidean algorithm as follows. We determine how often b is contained

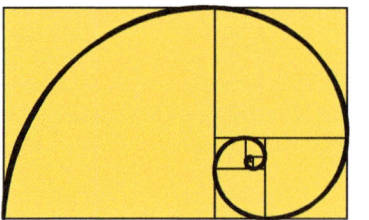

Fig. 2.47 Overlay. Quarter sectors

Fig. 2.48 Fractal

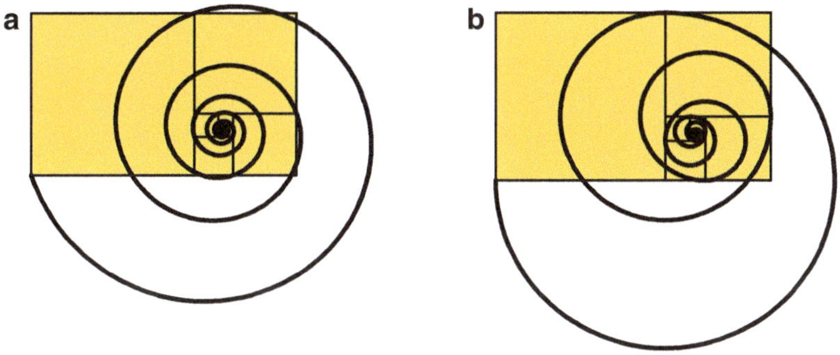

Fig. 2.49 Large Spirals to the Golden Rectangle

in a by dividing a by b and calculate the remainder. Then b is divided by the remainder, and the new remainder is determined. In each subsequent step, the old remainder is divided by the new remainder, and this continues until the division "goes on," that is, no remainder remains. The last non-zero remainder is then the greatest common divisor of a and b. Any further common divisor is also a divisor of this greatest common divisor. This method of dividing with remainder is called the Euclidean algorithm.

For the example $a = 42$ and $b = 15$ the result is obtained step by step:

$$42 = 2 \cdot 15 + 12 \qquad (2.17)$$

Fig. 2.50 Three-quarter circles
(▸ https://doi.org/10.1007/000-cat)

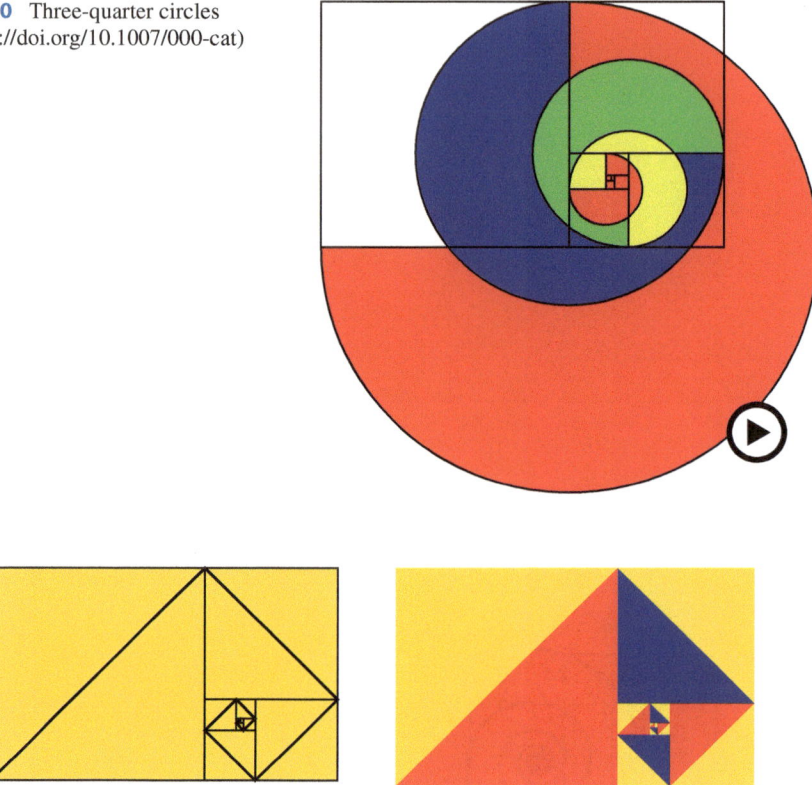

Fig. 2.51 Angular spiral in the Golden Rectangle

$$15 = 1 \cdot 12 + 3$$

$$12 = 4 \cdot 3 + 0$$

The greatest common divisor of 42 and 15 is therefore 3.

Geometric representation of the Euclidean algorithm

From a rectangle with side lengths a and b ($a \geq b$), squares with side length b are cut off as long as possible. Then a remaining rectangle remains, from which squares are again cut off and so on. The process continues until the decomposition into squares is complete, that is, no rectangle remains. The side of the smallest squares is then the greatest common divisor of a and b.

Figure 2.52 illustrates the procedure for $a = 42$ und and $b = 15$.

Since this smallest square side is a common measure (even the greatest common measure) of the original rectangle sides a and b, the entire rectangle can now be divided into squares of this size (Fig. 2.53).

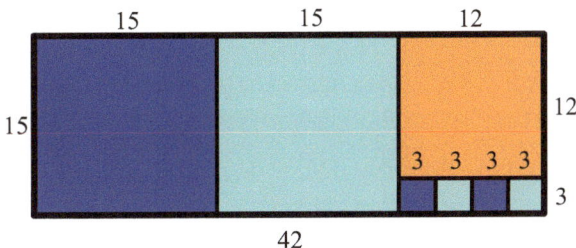

Fig. 2.52 Geometric representation of the Euclidean algorithm

Fig. 2.53 Division into squares

If now two lengths a and b have a common measure g, then $a = mg$ and $b = ng$ with $m, n \in \mathbb{N}$, and the ratio $b : a$ can be represented as a ratio of whole numbers $n : m$, thus as a rational number $\frac{n}{m}$. We then say, a and b are commensurable.

Application to the Golden Rectangle

If we apply the Euclidean algorithm to the Golden Rectangle with side lengths $a = \Phi$ and $b = 1$, the process never ends, as after cutting off a square, a rectangle similar to the original rectangle always remains. The side lengths Φ and 1 therefore have no common measure; the ratio $\Phi : 1 = \Phi$ cannot be expressed as an integer ratio, Φ is an irrational number. The first historical proof of incommensurability is probably by Hippasus of Metapontum in the 2nd quarter of the 5th century BC, using the number Φ, but with a different geometric consideration (see [34], p. 132).

2.3.6   Generalizations of the Golden Rectangle

We now examine rectangles where, after cutting off n squares, a rectangle similar to the original rectangle remains (Fig. 2.54 for $n = 3$).

The original rectangle has the length x and the width 1. From the similarity to the remaining rectangle, the condition for x is:

$$\frac{x}{1} = \frac{1}{x - n} \qquad (2.18)$$

This results in the quadratic equation

$$x^2 - nx - 1 = 0 \qquad (2.19)$$

Fig. 2.54 Remaining
rectangle similar to the
original rectangle

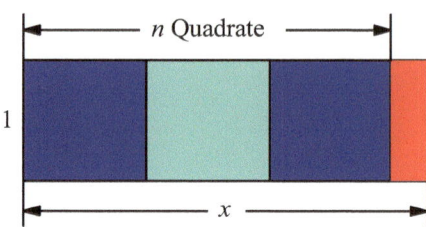

with the positive solution:

$$x = \frac{n + \sqrt{n^2 + 4}}{2} \tag{2.20}$$

For increasing n the rectangle becomes longer and longer.

The application of the Euclidean algorithm to the rectangles described here
leads, analogously to the considerations with the Golden Rectangle, to the conclu-
sion that the numbers of the form

$$x = \frac{n + \sqrt{n^2 + 4}}{2}, n \in \mathbb{N} \tag{2.21}$$

are irrational.

We now want to focus on the case $n = 2$, that is, the rectangle with side lengths
$1 + \sqrt{2}$ and 1 (cf. [24]). This rectangle is obtained by cutting off a square from a
paper in DIN format. The rectangle can be divided symmetrically into squares; the
diagonals support the subdivision process (Fig. 2.55). The diagonals intersect at
an angle of 45°. Therefore, four such rectangles can be stacked corner to corner to
form a regular octagon.

With the help of this subdivision, two intersecting point-symmetric spirals-
from quarter circles can be drawn or a figure from semicircles can be created (Fig.
2.56).

Fig. 2.55 Symmetrical subdivision. Octagon

Fig. 2.56 Double spirals and semicircles

Instead of placing the n squares next to each other, we can stack them on top of each other (Fig. 2.57 for $n = 3$). The rectangle can be divided into ever smaller squares in a spiral pattern.

With the designations of Fig. 2.57 the following results:

$$x = \frac{1 + \sqrt{1 + 4n^2}}{2n} \tag{2.22}$$

For increasing n the rectangle approaches a square.

2.3.7   The Tall Golden Rectangle

By the tall golden rectangle we understand the rectangle with the aspect ratio $\Phi : \frac{1}{\Phi}$ (Fig. 2.58). It looks quite long.

The aspect ratio of the tall golden rectangle can be written in different ways:

$$\Phi : \frac{1}{\Phi} = \Phi^2 : 1 = (1 + \Phi) : 1 \approx 2{,}618 : 1 \tag{2.23}$$

From the golden rectangle we obtain the tall golden rectangle as follows. We can either attach a square to a narrow side of the golden rectangle or attach a square to both long sides of a (standing upright) golden rectangle (Fig. 2.59).

Fig. 2.60 shows a division of a square with side length Φ into tall golden rectangles in landscape format and ordinary golden rectangles in portrait format.

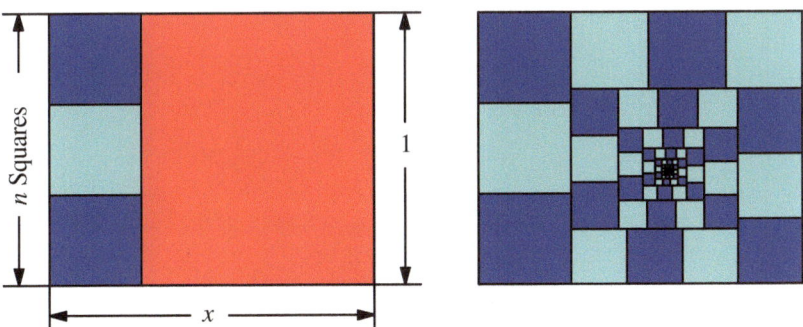

Fig. 2.57 Remaining rectangle similar to the original rectangle. Division

Fig. 2.58 Tall golden
rectangle

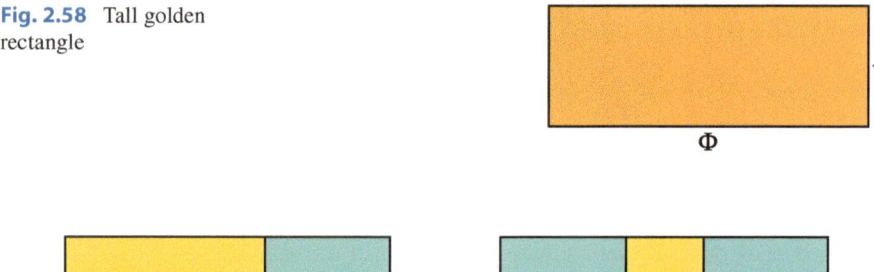

 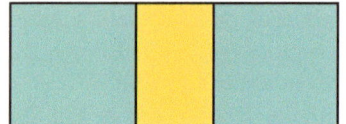

Fig. 2.59 Attaching squares

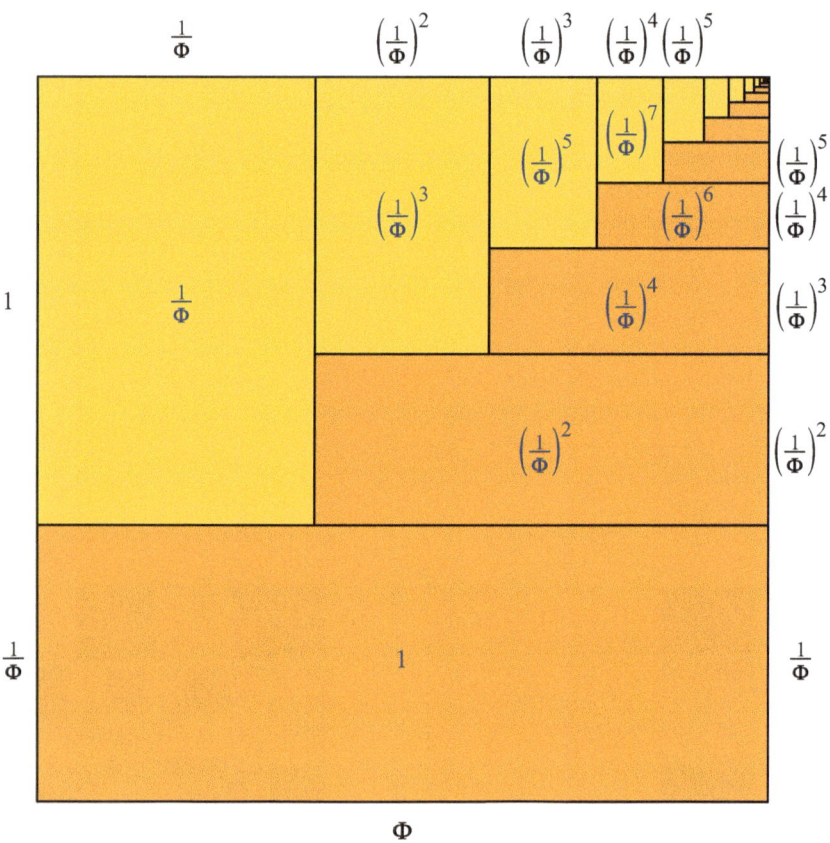

Fig. 2.60 Division of a square

The black measurements refer to the lengths of the routes, the blue ones to the areas. We read the relationships from this:

$$\Phi = \frac{1}{\Phi} + \left(\frac{1}{\Phi}\right)^2 + \left(\frac{1}{\Phi}\right)^3 + \cdots = \sum_{n=1}^{\infty} \left(\frac{1}{\Phi}\right)^n \qquad (2.24)$$

$$\Phi^2 = 1 + \frac{1}{\Phi} + \left(\frac{1}{\Phi}\right)^2 + \left(\frac{1}{\Phi}\right)^3 + \cdots = \sum_{n=0}^{\infty} \left(\frac{1}{\Phi}\right)^n \qquad (2.25)$$

We also find the high golden rectangle under the hyperbola. To do this, we first fit a square with one corner at the origin under the hyperbola $y = \frac{1}{x}$ (Fig. 2.61). To the right of it, we fit a second square.

The rectangle parallel to the axes with one corner at the origin and the diametrical corner in the upper right corner of the second square is a high golden rectangle.

In the right-angled triangle with the hypotenuse sections Φ and $\frac{1}{\Phi}$ (Fig. 2.62), the high golden rectangle formed from these sections has the same area as the height square, namely one (height theorem). This can be illustrated by a common decomposition.

We now try to show the area equality by common exhaustion with golden rectangles. After the first two steps, a figure remains which again consists of a high

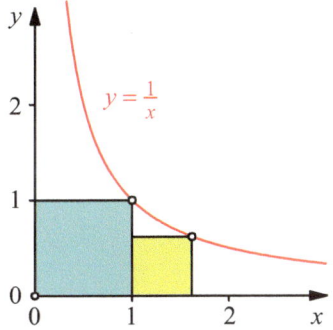 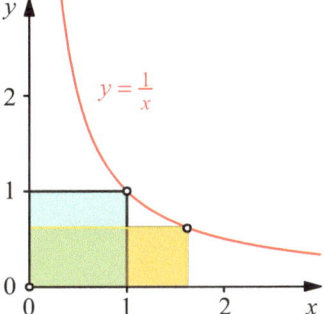

Fig. 2.61   Under the hyperbola

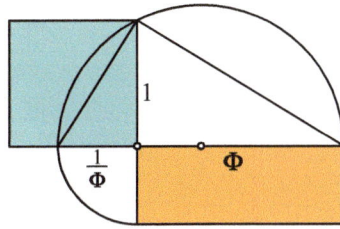 

Fig. 2.62   Area equality

 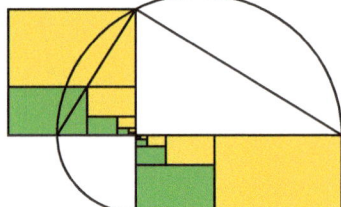

Fig. 2.63 Endless exhaustion

golden rectangle and a square (Fig. 2.63). So we are, so to speak, "back to square one" and can continue the process ad infinitum (forever).

2.4   Golden Polygons

In this section, we will get to know further figures in which, after cutting off suitable simple figures, a residual figure similar to the original figure remains.

2.4.1   The Golden Parallelogram

From the Golden Parallelogram with the side ratio of the Golden Section and the acute angle 60° two equilateral triangles can be cut off, so that the remaining parallelogram is again a Golden Parallelogram. The Golden Parallelogram can be divided into equilateral Triangles and with spirals provided (Fig. 2.64).

The Fig. 2.65 provides two generalizations of the Golden Parallelogram.

2.4.2   Golden Triangles

Already in the section about the regular pentagon, we encountered the acute and obtuse Golden Triangle (Fig. 2.26, 2.27 and 2.28) with the base angles of 72° and 36°.

Fig. 2.64 Division and spirals in the Golden Parallelogram

Fig. 2.65   Generalizations of the Golden Parallelogram

Fig. 2.66   Decomposition of
the acute Golden Triangle

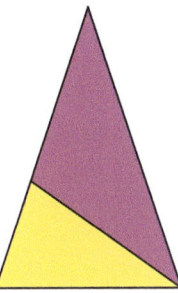

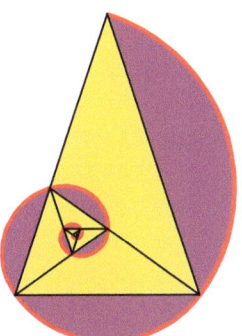

From the acute Golden Triangle, we can cut off an obtuse Golden Triangle, leaving an acute Golden Triangle (Fig. 2.66). This leads to a decomposition of the acute Golden Triangle into obtuse Golden Triangles when iterated. The decomposition can be provided with a spiral composed of circular arcs composite spiral.

Conversely, the obtuse Golden Triangle can be decomposed into acute Golden Triangles (Fig. 2.67).

In both cases, there is a similarity between the remaining triangle and the original triangle with the similarity factor $\frac{1}{\Phi}$.

A regular pentagon can be decomposed into acute and obtuse Golden Triangles as well as additional pentagons using the diagonals (Fig. 2.68).

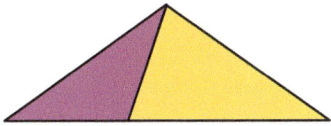

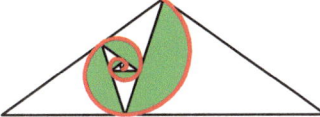

Fig. 2.67   Decomposition of the obtuse Golden Triangle

Fig. 2.68 Pentagon subdivisions

2.5 Golden Ellipses

Example 1: We compare the area of an ellipse with the semi-axes a and b with the area of the Thales circle over the foci F_1 and F_2 of the ellipse (Fig. 2.69, left). For which axis ratio $\frac{a}{b}$ do the ellipse and the Thales circle have the same area?

The ellipse has the area $ab\pi$; the half focal length is $\sqrt{a^2 - b^2}$. Since this is the radius of the Thales circle, the area of it is $(a^2 - b^2)\pi$. Equating the two areas yields $a^2 - b^2 = ab$, or after division by b^2:

$$\left(\frac{a}{b}\right)^2 - \frac{a}{b} - 1 = 0 \qquad (2.26)$$

Thus, $\frac{a}{b} =$ is Φ; the semi-axes of the ellipse are in the ratio of the golden section.

If we choose the radius of the Thales circle 1, then $a = \sqrt{\Phi}$ and $b = \frac{1}{\sqrt{\Phi}}$.

Example 2: In the grid (Fig. 2.69 right), we draw the ellipse with the grid focal points F_1 and F_2 through the grid point A. The long axis is divided by the focal

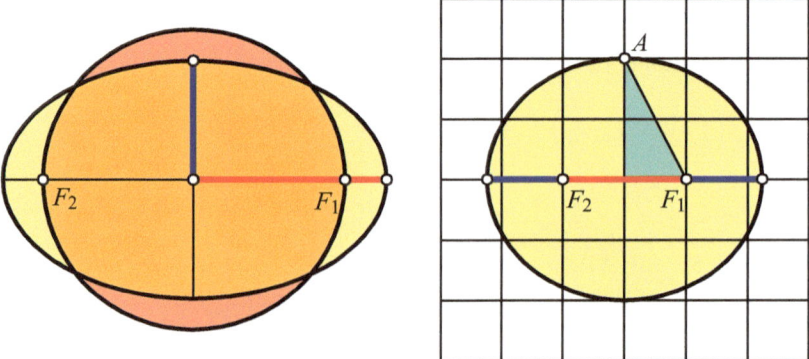

Fig. 2.69 Ellipse and circle with equal area. Ellipse in grid

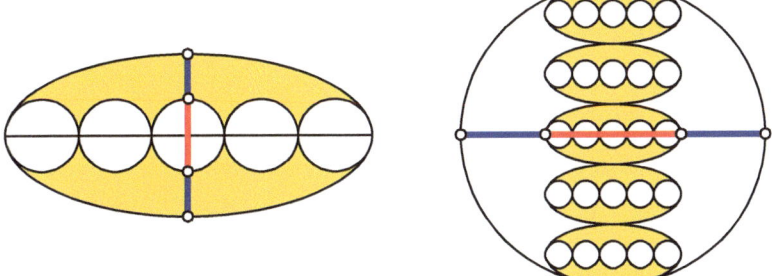

Fig. 2.70 Ellipse and curvature circles

point in the ratio of the golden section, in the order Minor-Major-Minor. The light blue triangle provides a clue to the proof.

Example 3: We start with five adjacent circles and place an ellipse around them so that the two outermost circles are the curvature circles at the sharp verticesof the ellipse (Fig. 2.70). The short semi-axis is divided by the middle circle in the ratio of the golden section, in the order Minor-Major-Minor.

We now stack five such ellipses on top of each other. The curvature circle at the upper blunt vertex of the top ellipse is also the curvature circle at the lower blunt vertex of the bottom ellipse. Here too, we find the golden section, again in the order Minor-Major-Minor.

2.6   Square Root of the Golden Ratio

2.6.1   Geometric Sequence

We are looking for a right-angled triangle, whose legs and the hypotenuse form a geometric sequence.

We set $a = 1$, $b = q$ and $c = q^2$. From the Pythagorean theorem follows:

$$1 + q^2 = q^4 \tag{2.27}$$

The positive solution of this biquadratic equation is $q = \sqrt{\Phi} \approx 1{,}272$. Thus, $a = 1$, $b = \sqrt{\Phi}$ and $c = \Phi$ (Fig. 2.71). The leg b is the geometric mean of a and c.

The longer hypotenuse segment has the length 1 like the leg a. The two hypotenuse segments are in the ratio of Major to Minor. The triangle height is the reciprocal of the square root of the Golden Ratio.

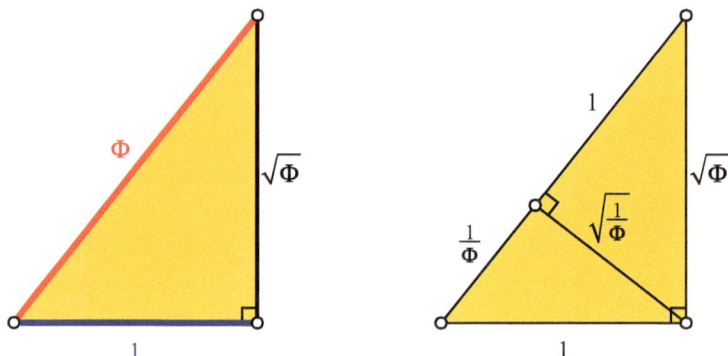

Fig. 2.71 Square root of the Golden Ratio

2.6.2 Golden Staircase

The Golden Staircase is a non-convex hexagon that can be **divided** into two similar, but not congruent sub-hexagons (Fig. 2.72) ([23], p. 53).

The side lengths are exactly the powers $\sqrt{\Phi^k}$ with $k = 0, \ldots, 5$. These numbers form a geometric sequence, but they are not arranged in order on the perimeter of the Golden Staircase.

The area ratio of the two subfigures is $\Phi : 1$.

Since the subfigures are also Golden Staircases, they can also be divided into two Golden Staircases (Fig. 2.73). The subfigures are colored in order of size in red, blue, yellow, green, magenta, and purple.

Table 2.1 provides an overview of the frequencies of equally sized subfigures. We obtain the binomial coefficients.

We can assemble Golden Staircases into a spiral (Fig. 2.74).

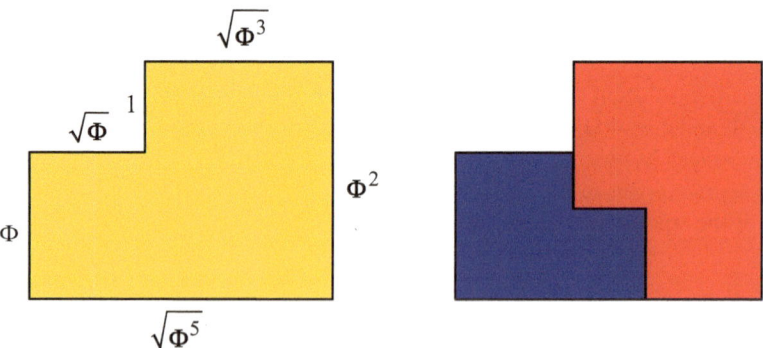

Fig. 2.72 Golden Staircase. Decomposition

Fig. 2.73 Decompositions

Table 2.1 The frequencies appear as binomial coefficients

1					
1	1				
1	2	1			
1	3	3	1		
1	4	6	4	1	
1	5	10	10	5	1
frequencies					

2.6.3 Double Spiral

The double spiral Fig. 2.75) consists of two point-symmetrically located spirals constructed from equilateral triangles. The outer profile of one spiral touches the inner profile of the other and vice versa.

To calculate the length reduction factor f when transitioning from one triangle to the next, we refer to Fig. 2.75:

$$\frac{1}{2}\sqrt{3}\left(f + f^2\right) = \frac{1}{2}\sqrt{3}\left(f^3 + f^4 + f^5 + f^6\right) \qquad (2.28)$$

Since the trivial solution $f = 0$ is not relevant to us, we can divide the equation by f and simplify to:

$$1 + f = (1 + f)\left(f^2 + f^4\right) \qquad (2.29)$$

Another solution is thus $f = -1$. This is also not relevant to us, so we can divide by the corresponding linear factor. This results in:

$$1 = f^2 + f^4 \qquad (2.30)$$

Fig. 2.74 Staircase spiral

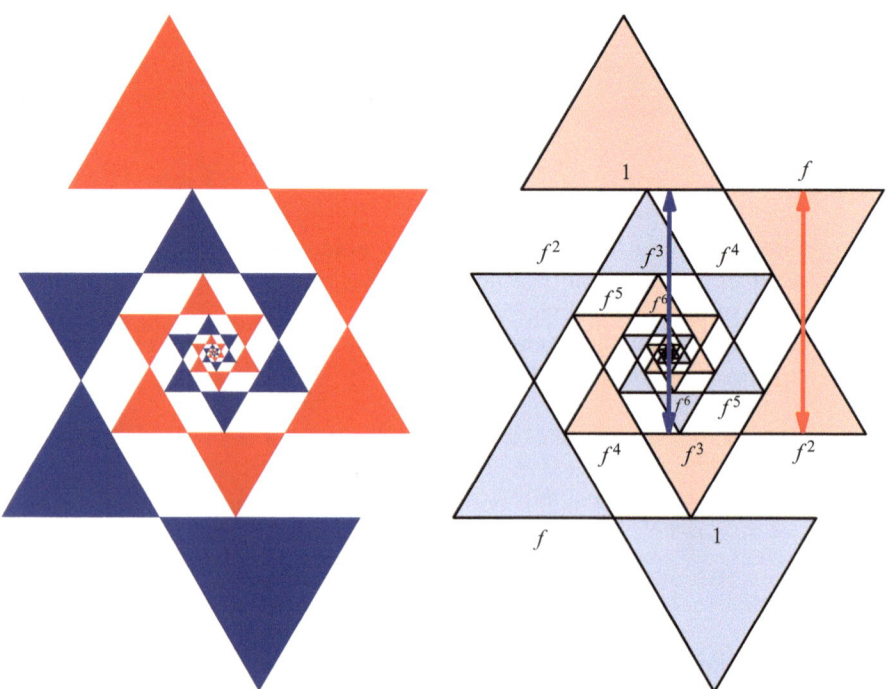

Fig. 2.75 Double spiral. Dimensions

From this biquadratic equation we obtain:

$$f^2 = \frac{-1 \pm \sqrt{5}}{2} \tag{2.31}$$

For real solutions for f only the plus solution is relevant, so $f^2 = \frac{1}{\Phi}$ and thus $f = \pm\sqrt{\frac{1}{\Phi}}$. Again, only the positive solution is relevant, so we finally have:

$$f = \sqrt{\frac{1}{\Phi}} \approx 0{,}7865 \tag{2.32}$$

The transition factor is thus the square root of the golden ratio.

2.7   Area Divisions

2.7.1   Dividing a Square

A square is divided into five equal parts (Fig. 2.76). The Golden Ratio appears in this process.

2.7.2   Extending Sides

In any triangle, we extend two sides in the ratio of the Golden Section (Fig. 2.77).

Now we fit in a supplementary triangle (yellow in Fig. 2.78).

This yellow supplementary triangle has the same area as the starting triangle. For the proof, we work with the labels of Fig. 2.78.

We calculate the area F of the green triangle with:

$$F = \frac{1}{2}ab \sin(\gamma) \tag{2.33}$$

Because of $\sin(180° - \gamma) = \sin(\gamma)$ we get correspondingly for the area G of the yellow supplementary triangle:

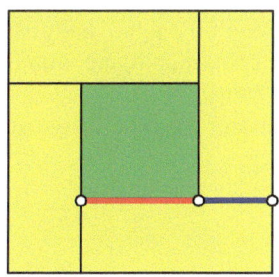 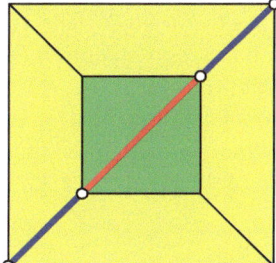

Fig. 2.76 Division into five parts

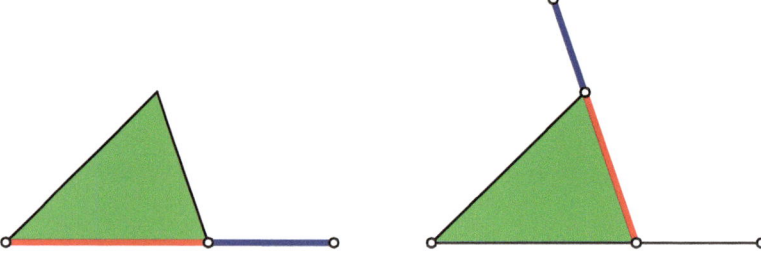

Fig. 2.77 Extending sides

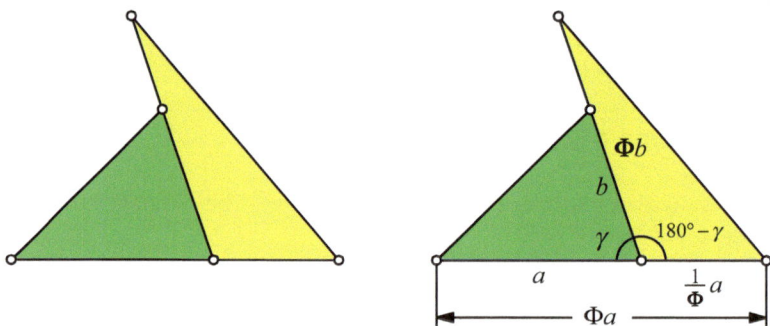

Fig. 2.78 Supplementary triangle

$$G = \frac{1}{2}\frac{1}{\Phi}a\Phi b \sin\left(180° - \gamma\right) = \frac{1}{2}ab\sin\left(\gamma\right) \qquad (2.34)$$

So, the two areas are equal in size.

Fig. 2.79 provides a common division of the two triangles. Each piece in the green starting triangle also appears in the yellow supplementary triangle and vice versa. The corresponding pieces can be transformed into each other either by translation or point reflection.

We can extend all three sides of the green starting triangle and draw three supplementary triangles (Fig. 2.80). Each supplementary triangle is equal in area to the green starting triangle. Therefore, the total area of the figure has quadrupled.

Through repeated repetition of the process, a spiral-like figure with exponential area growth is created.

With a suitable diagonal, it can be shown that in the case of the square (Fig. 2.81), the total figure is three times as large in terms of area as the starting square. In the case of the regular hexagon, the total figure is twice as large.

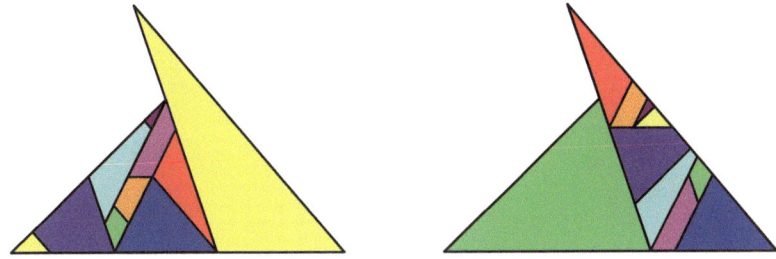

Fig. 2.79 Common division

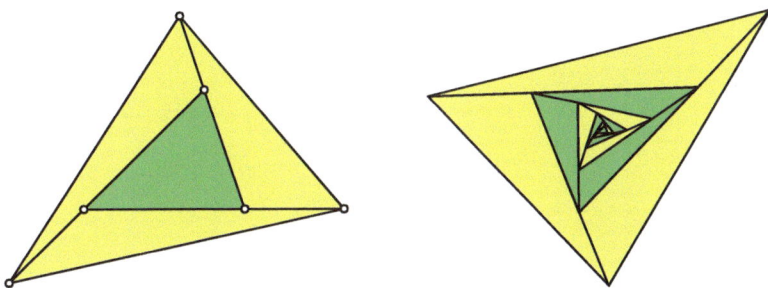

Fig. 2.80 Three supplementary triangles. Repetition

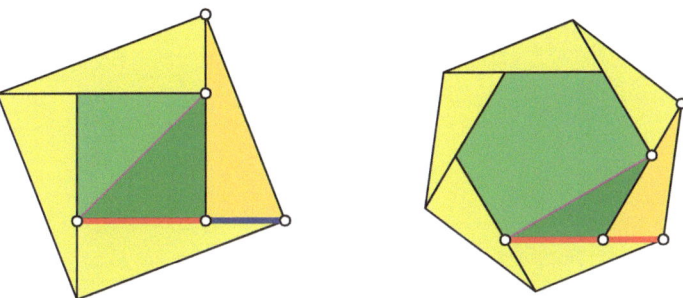

Fig. 2.81 Square and hexagon

Golden Angles

3

3.1 Examples

3.1.1 Golden Pie

In a 36° sector (one tenth of the whole pie), the perpendicular bisector of one radial side divides the other in the ratio of the Golden Section (Fig. 3.1).

Of course, this does not provide a construction method for the Golden Section, because for the construction of a 36° angle, we already need the Golden Section in some way.

3.1.2 Cosine and Tangent

Figure 3.2 shows the graphs of the cosine function and of the tangent function. Where and at what angle do the two curves intersect? (see [29])

To calculate the intersection point we solve the equation:

$$\cos(x) = \tan(x) \tag{3.1}$$

Transformation leads to the equation:

$$1 - \sin^2(x) = \sin(x) \tag{3.2}$$

This quadratic equation for $\sin(x)$ has the solution relevant to us in radian measure:

Supplementary Information The electronic version of this chapter contains additional material, which can be accessed via the following link https://doi.org/10.1007/978-3-662-69890-7_3. The videos can be played by clicking on the DOI link in the legend of a corresponding figure, or by scanning this link with the SN More Media App.

Fig. 3.1 Golden Pie

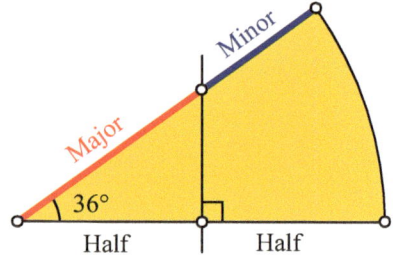

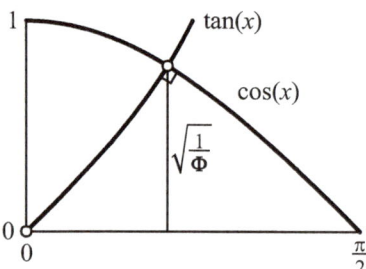

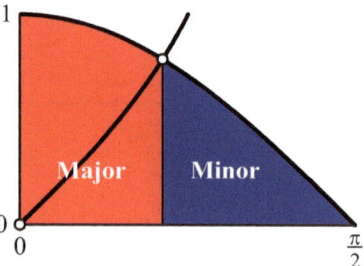

Fig. 3.2 Cosine and Tangent

$$\sin(x) = \frac{1}{\Phi} \Rightarrow x = \arcsin\left(\frac{1}{\Phi}\right) \approx 0.6662 \tag{3.3}$$

Further, we have:

$$\cos\left(\arcsin\left(\frac{1}{\Phi}\right)\right) = \sqrt{1 - \sin^2\left(\arcsin\left(\frac{1}{\Phi}\right)\right)} = \sqrt{1 - \frac{1}{\Phi^2}} = \sqrt{\frac{1}{\Phi}} \approx 0.7862 \tag{3.4}$$

For the intersection point, we thus obtain the coordinates:

$$\left(\arcsin\left(\frac{1}{\Phi}\right), \sqrt{\frac{1}{\Phi}}\right) \approx (0.6662, 0.7862) \tag{3.5}$$

At the intersection point, the graph of the cosine function has the slope $-\frac{1}{\Phi}$ and the graph of the tangent function has the slope Φ. The product of the two slopes is -1, so the two function graphs intersect orthogonally.

Further, for the red marked area piece, we get:

$$\int_0^{\arcsin\left(\frac{1}{\Phi}\right)} \cos(x)dx = \frac{1}{\Phi} \tag{3.6}$$

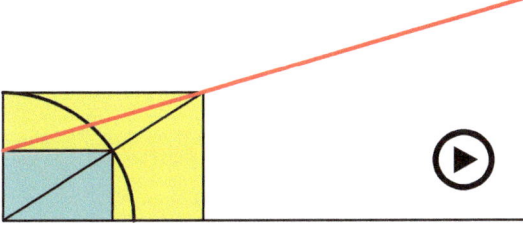

Fig. 3.3 Kinematics (▶ https://doi.org/10.1007/000-cay)

For the blue marked area piece, we get accordingly:

$$\int_{\arcsin\left(\frac{1}{\Phi}\right)}^{\frac{\pi}{2}} \cos(x)dx = \left(\frac{1}{\Phi}\right)^2 \tag{3.7}$$

The two area pieces are thus in the ratio of Major to Minor. We have an area ratio in the Golden Section.

3.1.3 The Steepest Line

Fig. 3.3 shows a mechanism with two similar rectangles, a quarter circle, and a straight line. This line runs through the upper corner of each rectangle.

The red line is initially horizontal, then becomes steeper, but then the slope decreases again and at the end the line is horizontal again. In which position is the line the steepest?

For processing, we use the designations of Fig. 3.4.

For the slope $m(t)$ of the red line we read:

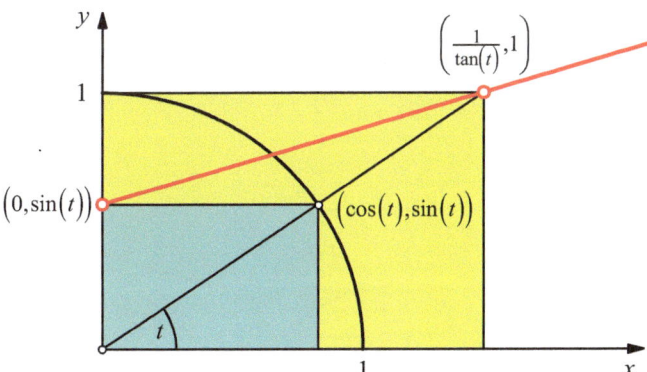

Fig. 3.4 Designations

$$m(t) = \frac{1 - \sin(t)}{\frac{1}{\tan(t)}} = \tan(t)(1 - \sin(t)) \tag{3.8}$$

The optimization condition

$$\frac{dm}{dt}(t) = 0 \tag{3.9}$$

leads to the equation:

$$\sin^3(t) - 2\sin(t) + 1 = 0 \tag{3.10}$$

This cubic equation has for $\sin(t)$ the three solutions $\sin(t_1) = 1$, $\sin(t_2) = \frac{1}{\Phi}$ and $\sin(t_3) = -\Phi$. The solution relevant to us is $\sin(t_2) = \frac{1}{\Phi}$.

The parameter value t (in radian measure) corresponding to the steepest red line is therefore:

$$t = \arcsin\left(\frac{1}{\Phi}\right) \approx 0.6662 \tag{3.11}$$

In degrees, this is $38.173°$; however, this is not the slope angle of the red line, but the slope angle of the common diagonal of the two rectangles.

The corresponding maximum slope of the red line is $m \approx 0.3003$. Figure 3.5 shows the corresponding line. The slope angle is $16.714°$.

For the steepest line, the two rectangles have a side ratio:

$$1 : \sqrt{\Phi} \approx 1 : 1.272 \tag{3.12}$$

3.1.4   Rectangle in the Semicircle

In a kinematic game in the semicircle (Fig. 3.6), angles appear in the context of the numbers $\sqrt{2}$ (DIN format), $\sqrt{3}$ (regular hexagon) and the Golden Ratio (and thus the number $\sqrt{5}$).

The two green lines rotate synchronously. They always form a right angle. We examine three special stations of the process. These are the stops in Fig. 3.6.

At the first intermediate stop (Fig. 3.7), we obtain the condition for the angle t_1:

Fig. 3.5 Steepest line

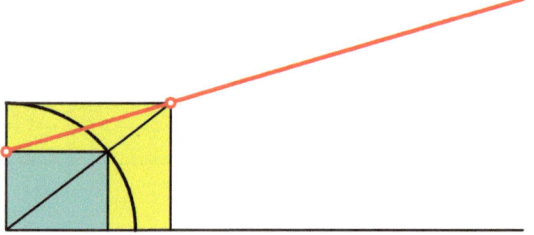

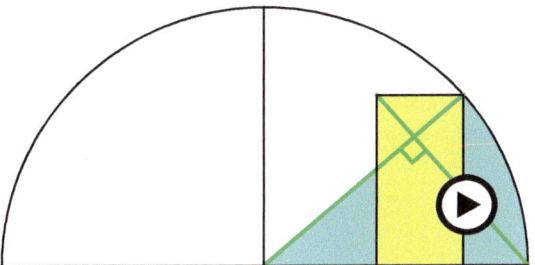

Fig. 3.6 Rectangle in the semicircle (▶ https://doi.org/10.1007/000-caw)

Fig. 3.7 At the first
intermediate stop

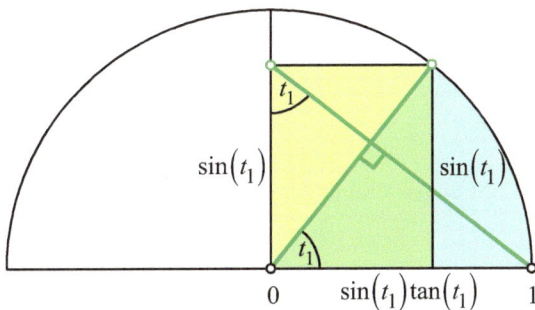

$$\sin (t_1) \tan (t_1) = 1 \tag{3.13}$$

This equation can be transformed into:

$$\frac{\sin^2(t_1)}{\sqrt{1 - \sin^2(t_1)}} = 1 \tag{3.14}$$

From this, the biquadratic equation for $\sin (t_1)$ results:

$$\sin^4(t_1) + \sin^2(t_1) - 1 = 0 \tag{3.15}$$

The solution relevant to us is:

$$\sin (t_1) = \sqrt{\frac{1}{\Phi}} \tag{3.16}$$

Thus, in radian measure:

$$t_1 = \arcsin \left(\sqrt{\frac{1}{\Phi}} \right) \approx 0.905 \tag{3.17}$$

In degree measure, we get $t_1 \approx 51.827°$.

Fig. 3.8 Second intermediate stop

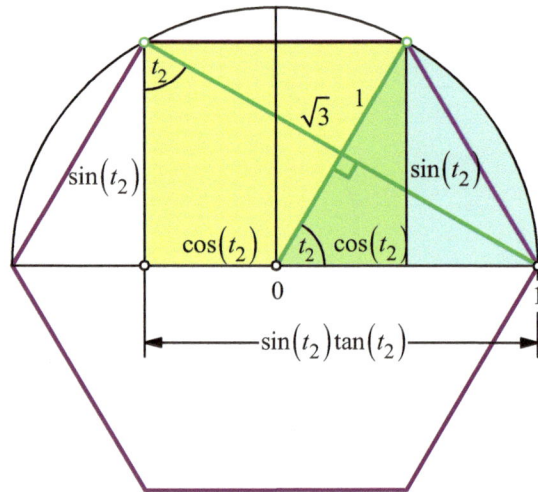

At the second intermediate stop (Fig. 3.8), we obtain the condition:

$$\sin (t_2) \tan (t_2) = 1 + \cos (t_2) \tag{3.18}$$

We can transform this equation into:

$$2\cos^2 (t_2) + \cos (t_2) - 1 = 0 \tag{3.19}$$

The solution relevant to us is $\cos (t_2) = \frac{1}{2}$. Therefore:

$$t_2 = \arccos \left(\frac{1}{2} \right) = \frac{\pi}{3} = 60° \tag{3.20}$$

The figure can be fitted into a regular hexagon. The length ratio of the two green lines is $\sqrt{3} : 1$.

For the terminal station (Fig. 3.9), we obtain the condition:

Fig. 3.9 Terminal station

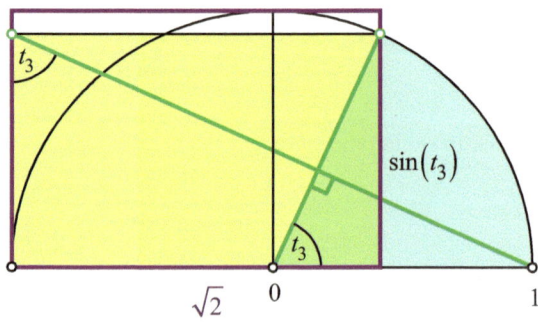

$$\sin(t_3)\tan(t_3) = 2 \tag{3.21}$$

This can be transformed to:

$$\cos^2(t_3) + 2\cos(t_3) - 1 = 0 \tag{3.22}$$

The solution relevant to us is $\cos(t_3) = \sqrt{2} - 1$. Thus, we have:

$$t_3 = \arccos\left(\sqrt{2} - 1\right) \approx 1.1437 \approx 65.5302° \tag{3.23}$$

The figure can be fitted into a rectangle in DIN format (aspect ratio $\sqrt{2} : 1$).

Between the first and the second stop, there is a point t_Q, where the yellow rectangle is a square (Fig. 3.10).

For this square, we obtain the condition:

$$\sin(t_Q) = \sin(t_Q)\tan(t_Q) - \left(1 - \cos(t_Q)\right) \tag{3.24}$$

This condition leads to the equation:

$$\sin^4(t_Q) + 2\sin^2(t_Q) - 2\sin(t_Q) = 0 \tag{3.25}$$

We now have not just a quadratic or biquadratic equation, but a fourth-degree equation. The solution relevant to us is:

$$t_Q = \arcsin\left(\frac{\sqrt[3]{19 + 3\sqrt{33}}}{3} + \frac{4}{3\sqrt[3]{19 + 3\sqrt{33}}} - \frac{2}{3}\right) \approx 0.9960 \approx 57.0649° \tag{3.26}$$

3.1.5 A Trigonometric Relationship

In Sects. 3.1.2 and 3.1.3 we encountered the angle:

$$\arcsin\left(\frac{1}{\Phi}\right) \approx 38.173° \tag{3.27}$$

Fig. 3.10 Square

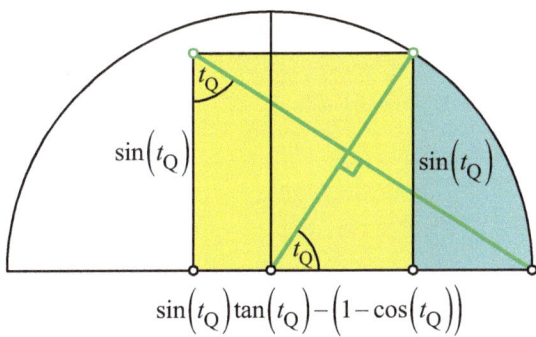

In Sect. 3.1.4 we found the angle:

$$\arcsin\left(\sqrt{\frac{1}{\Phi}}\right) \approx 51.827°$$

We suspect that these two angles add up to 90°:

$$\arcsin\left(\frac{1}{\Phi}\right) + \arcsin\left(\sqrt{\frac{1}{\Phi}}\right) = 90° \qquad (3.28)$$

We can verify the consistency of this formula in the right-angled triangle with the catheti $\sqrt{\frac{1}{\Phi}}$ and $\frac{1}{\Phi}$ (Fig. 3.11).

For the hypotenuse of this triangle we obtain:

$$c^2 = \left(\frac{1}{\Phi}\right)^2 + \left(\sqrt{\frac{1}{\Phi}}\right)^2 = \left(\frac{1}{\Phi}\right)^2 + \frac{1}{\Phi} = 1 \qquad (3.29)$$

Thus, it is:

$$\begin{aligned} \alpha &= \arcsin\left(\frac{1}{\Phi}\right) \\ \beta &= \arcsin\left(\sqrt{\frac{1}{\Phi}}\right) \end{aligned} \qquad (3.30)$$

In a right-angled triangle, however, the sum of the two acute angles is 90°.

The angle β can also be written in the simpler form $\beta = \arctan\left(\sqrt{\Phi}\right) \approx 51.827°$.

The Great Pyramid of Cheops is said to have been built with a base length of 440 cubits (about 230.83 m) and a height of 280 cubits (about 146.60 m). This results in an angle of inclination of the side surfaces $\arctan\left(\frac{14}{11}\right) \approx 51.84°$. This is approximately the angle $\beta = \arctan\left(\sqrt{\Phi}\right) \approx 51.827°$. Therefore, $\frac{14}{11} \approx 1.273$ is an approximation of $\sqrt{\Phi} \approx 1.272$. However, there is a fundamental difference between $\frac{14}{11}$ and $\sqrt{\Phi}$. The number $\frac{14}{11}$ is a rational number, whereas $\sqrt{\Phi}$ is irrational. Therefore, it is not meaningful to say that the Golden Ratio was used in the construction of the Great Pyramid of Cheops.

Fig. 3.11 Proof triangle. Pyramid

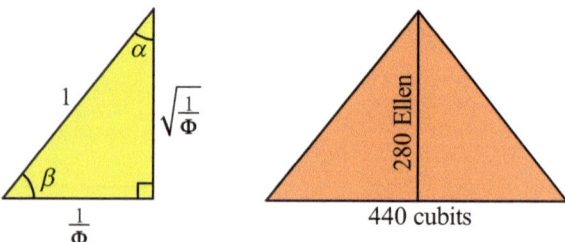

Fig. 3.12 Trigonometry in
the acute Golden Triangle

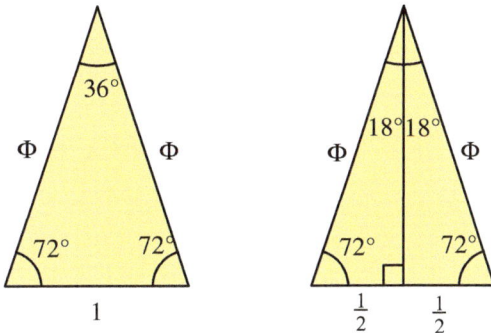

3.2   Golden Trigonometry

3.2.1   Formulas and Examples

From the proportions of the acute Golden Triangle (Fig. 3.12), we first obtain the
following relationships:

$$
\begin{aligned}
\cos(18°) &= \frac{\sqrt{2+\Phi}}{2} \\
\sin(18°) &= \frac{1}{2\Phi} = \frac{\Phi-1}{2} \\
\tan(18°) &= \frac{\Phi-1}{\sqrt{2+\Phi}}
\end{aligned}
\tag{3.31}
$$

With the help of the addition theorems, this results in the formulas of Tab. 3.1
Some examples on this topic:

$$
\begin{aligned}
\frac{\sin(66°)-\sin(6°)}{\cos(60°)} &= \Phi \\
\frac{\sin(78°)-\sin(42°)}{\sin(30°)} &= \frac{1}{\Phi} \\
\sin(18°)\cos(36°) &= \frac{1}{4}
\end{aligned}
\tag{3.32}
$$

Examples with arcus functions:

Tab. 3.1 Trigonometry

	18°	36°	54°	72°
cos	$\frac{\sqrt{2+\Phi}}{2}$	$\frac{\Phi}{2}$	$\frac{\sqrt{3-\Phi}}{2}$	$\frac{\Phi-1}{2}$
sin	$\frac{\Phi-1}{2}$	$\frac{\sqrt{3-\Phi}}{2}$	$\frac{\Phi}{2}$	$\frac{\sqrt{2+\Phi}}{2}$
tan	$\frac{\Phi-1}{\sqrt{2+\Phi}}$	$\frac{\sqrt{3-\Phi}}{\Phi}$	$\frac{\Phi}{\sqrt{3-\Phi}}$	$\frac{\sqrt{2+\Phi}}{\Phi-1}$

Trigonometry

$$\begin{aligned}
\Phi &= 1 + \tan\left(\tfrac{1}{2}\arctan(2)\right) \\
\Phi &= 2 - \tan\left(\tfrac{1}{2}\arcsin\left(\tfrac{2}{3}\right)\right) \\
\Phi &= \tan\left(90° - \tfrac{1}{2}\arctan(2)\right) \\
\tfrac{1}{\Phi} &= \tan\left(\tfrac{1}{2}\arctan(2)\right) \\
\tfrac{1}{\Phi} &= 1 - \tan\left(\tfrac{1}{2}\arcsin\left(\tfrac{2}{3}\right)\right) \\
\left(\tfrac{1}{\Phi}\right)^2 &= \tan\left(\tfrac{1}{2}\arcsin\left(\tfrac{2}{3}\right)\right)
\end{aligned} \tag{3.33}$$

3.2.2 Fourier and the Golden Ratio

Jean Baptiste Joseph Fourier (1768–1830) showed that periodic functions can be approximated by trigonometric functions. In particular, a function with the period length 2π can be approximated by a sum of the form

$$f(t) = \frac{a_0}{2} + \sum_{n=1}^{N}\left(a_n\cos(nt) + \sin(nt)\right) \tag{3.34}$$

with constant coefficients a_0 and $a_n, b_n (n = 1\dots N)$ approximated can be.

For example, the signal function (blue in Fig. 3.13) can be approximated by the sum

$$f(t) = \frac{8}{\pi}\left(\sin(t) + \frac{1}{3}\sin(3t) + \frac{1}{5}\sin(5t) + \cdots\right). \tag{3.35}$$

We now study some intersections with the special functions $\cos(nt)$ and $\sin(nt)$.

First, we look for the intersections of the two function graphs of $\cos(2t)$ and $\cos(3t)$ for $t \in [0, 2\pi]$ (Fig. 3.14). There are six intersections, which are evenly distributed with respect to t. This can be verified with some calculation using the addition theorems.

For the intersections, we obtain the coordinates of Tab. 3.2.

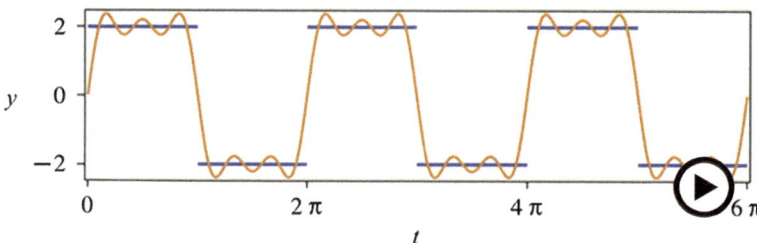

Fig. 3.13 Fourier Approximation (▶ https://doi.org/10.1007/000-cax)

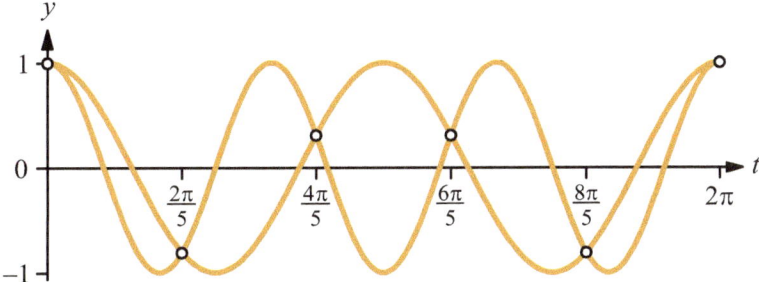

Fig. 3.14 Evenly distributed intersections

Tab. 3.2 Intersection coordinates

t	0	$\frac{2\pi}{5}$	$\frac{4\pi}{5}$	$\frac{6\pi}{5}$	$\frac{8\pi}{5}$	2π
y	1	$-\frac{\Phi}{2}$	$\frac{1}{2\Phi}$	$\frac{1}{2\Phi}$	$-\frac{\Phi}{2}$	1

Intersection coordinates

The Golden Ratio appears. However, this was to be expected, as the division of the full angle into five belongs to the regular pentagon.

In Fig. 3.15, the graphs of the functions $\cos(nt)$ for $n = 1 \dots 20$ are shown.

We soon see white ghost curves, which do not actually exist. However, they seem to fit into the trigonometric system.

Furthermore, we see certain passages obligés, i.e., points through which the curves must pass. Thus, the start and end are always at the height 1. In the middle, i.e., at $t = \pi$, only the top or bottom is possible, i.e., the heights 1 and -1. At the fifth positions (Fig. 3.16), only the three heights 1, $\frac{1}{2\Phi} \approx 0.309$ and $-\frac{\Phi}{2} \approx -0.809$ are possible. Between these three heights, we have the ratio of the Golden Section.

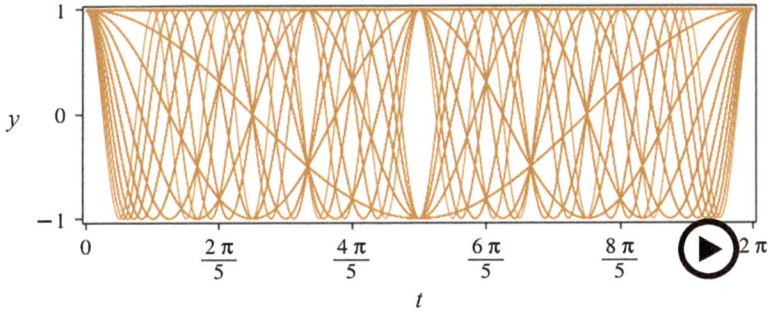

Fig. 3.15 Cosine functions. Ghost curves (▶ https://doi.org/10.1007/000-cav)

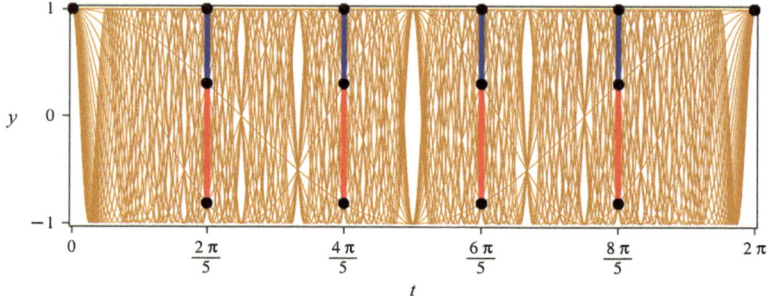

Fig. 3.16 Major and Minor

Fig. 3.17 Rosette with propeller and triangle (▶ https://doi.org/10.1007/000-caz)

3.2.3 Rosettes

When the function $r(t) = \cos(5t)$ is represented in polar coordinates, a rosette with five leaves is created (Fig. 3.17).

A propeller and an equilateral triangle can be fitted into this rosette, so that the ends of the propeller and the corners of the triangle lie on the edge of the rosette. This even works when we rotate the three figures each around their center. The rotations are at different speeds, but all have the same direction of rotation.

This figure can be expanded into a gear chain (Fig. 3.18).

A pink cloverleaf drives the propeller, this drives the five-leaf rosette (now we are at the golden ratio), this drives the equilateral triangle, this drives a seven-leaf rosette, this drives a square, this drives a nine-leaf rosette, this drives a regular pentagon (now we are back at the golden ratio) and the pentagon drives an eleven-leaf rosette. We see how it continues. The rotation speeds decrease harmoniously.

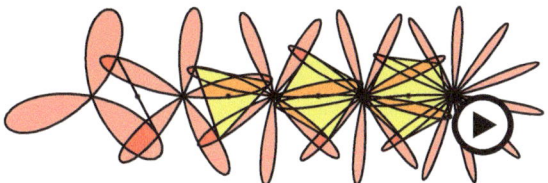

Fig. 3.18   Gear (▶ https://doi.org/10.1007/000-cb0)

Folding and Cutting

4.1 Paper Strip Construction of the Regular Pentagon

4.1.1 From Strip to Knot

We attempt to create a simple knot from a paper strip about 2 cm wide (Fig. 4.1). The strips can be cut out of ordinary printer paper in DIN A4 format.

Careful tightening and flattening results in a regular pentagon with two tails (cf. [14, 30, 31, 40, 41]).

If we use a strip of transparent paper and bend one of the two tails back, a star with five points appears inside the pentagon, a pentagram (Fig. 4.2).

By undoing the knot, we obtain a strip with diagonal fold lines (Fig. 4.3). The fold lines form an angle of $72°$ with the edge of the strip. This is one fifth of $360°$.

Regular pentagons can be fitted between the diagonal fold lines (Fig. 4.4).

4.1.2 Pre-Folding

We now attempt to pre-fold the strip with the diagonal fold lines in order to subsequently knot it into a pentagon.

In the following illustrations, it is assumed that the strip is two-colored, yellow on the front and light blue on the back (Fig. 4.5). Strips of this kind can be cut out of one-sided printed gift or cabinet paper. Of course, in practice, one can still work with a single-colored strip. The two colors are only used here to clearly illustrate the folding process.

Supplementary Information The electronic version of this chapter contains additional material, which can be accessed via the following link https://doi.org/10.1007/978-3-662-69890-7_4. The videos can be played by clicking on the DOI link in the legend of a corresponding figure, or by scanning this link with the SN More Media App.

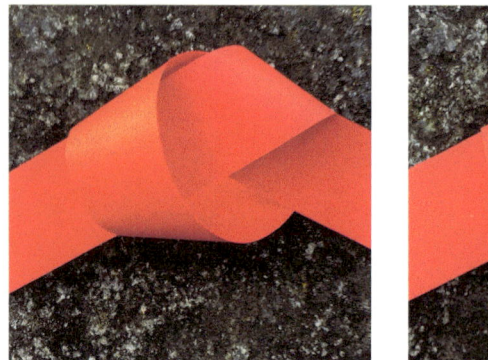

Fig. 4.1 Loose knot and tightened knot made from a paper strip

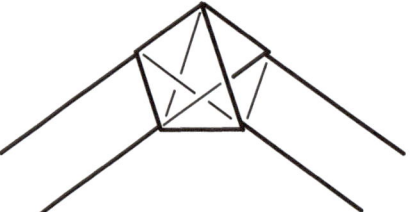
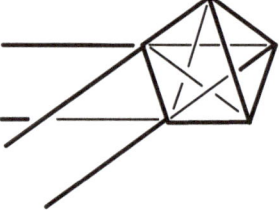

Fig. 4.2 Pentagon and Pentagram

Fig. 4.3 Diagonal fold lines

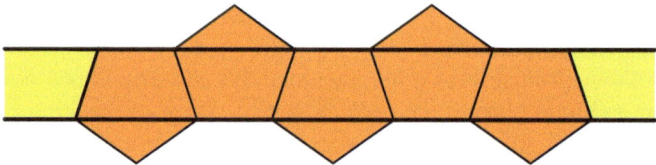

Fig. 4.4 Fitted Pentagons

Fig. 4.5 Strip with two colors

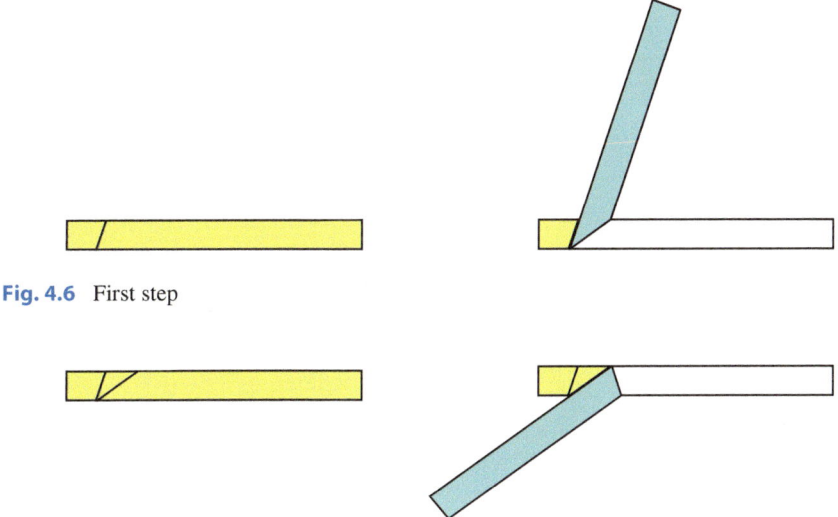

Fig. 4.6 First step

Fig. 4.7 Second step

We now mark the first fold line at the left end with an angle of 72° (Fig. 4.6). We will see that we only need to draw this angle by hand with the protractor for the first time, afterwards we get the angles automatically with the folding process. Furthermore, we fold the right part of the strip up so that the lower edge of the paper comes to lie just next to the red marked first fold line. In theory, the lower edge of the paper should even come exactly on the first fold line, but due to the thickness of the paper, we need to leave some leeway.

Now we fold back the part that was folded upwards (Fig. 4.7). The new fold line is the bisector of the 72° angle. So now we have angles of 36°. Accordingly, we also have an angle of 36° at the top edge of the paper and a supplementary angle of $180°-36° = 144°$.

Now we fold the right part of the strip down so that the top edge of the paper comes to lie just below the new fold line.

We fold back again (Fig. 4.8). The newly created fold line is the bisector of the 144° angle. So again we have 72° angles, without having to use the set square. The situation is now such that we could fit a regular pentagon. The long fold line then becomes a diagonal of the pentagon.

Now we fold the right part of the strip down so that the top edge of the paper comes to lie just next to the new fold line.

Fig. 4.9 shows the next step.

After folding back, we can now fit two pentagons (Fig. 4.10).

And so we continue working until we finally have space for five pentagons (Fig. 4.11).

Figure 4.12 shows the complete folding process.

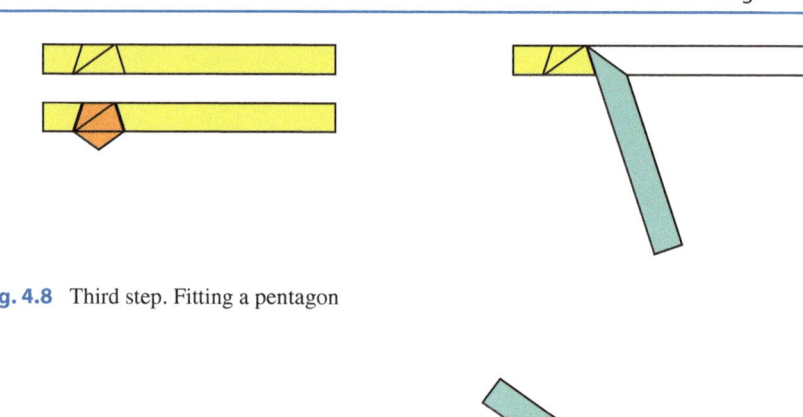

Fig. 4.8   Third step. Fitting a pentagon

Fig. 4.9   Fourth step

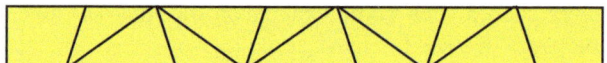

Fig. 4.10   Two pentagons can be fitted

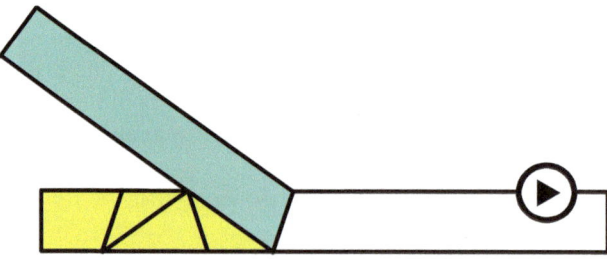

Fig. 4.11   Done

Fig. 4.12   Folding process (▶ https://doi.org/10.1007/000-cb4)

Fig. 4.13 Pre-folded strips

With this pre-folded strip, we can now elegantly construct the pentagonal knot build (Fig. 4.13).

4.1.3   A Self-Correcting Algorithm

Our folding process is somewhat boring in the long run because we have to perform the same step over and over again. A procedure in which the same step has to be repeated multiple times is called an algorithm. Because of the monotonous repetition of the same step, algorithms are suitable for mental meditations or for the use of a computer.

Our algorithm is self-correcting. Let's assume that we made a mistake when drawing the first angle by one degree, i.e., we drew an angle of $73°$ instead of $72°$ (Fig. 4.14). With the first angle halving by folding, we then get $36.5°$ instead of $36°$. The error is therefore still $0.5°$. When halving the supplementary angle of $143.5°$, we get $71.75°$. The error to the target value of $72°$ is therefore still $0.25°$, i.e., a quarter of the starting error. With each step, the error is halved and therefore tends towards zero.

This also means that we wouldn't actually need the set square to measure the first angle. We can start with any angle. With repeated execution of the algorithm, the angle increasingly approaches the value of $72°$, so that we soon have a practically useful approximation.

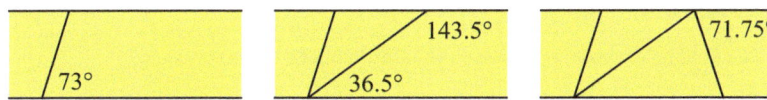

Fig. 4.14 Error becomes smaller and smaller

Fig. 4.15 Major and Minor.
Large Pentagon

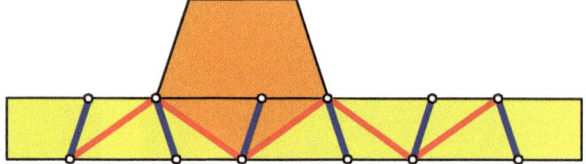

4.1.4 Another Pentagon

For the pentagonal node from our strip, we only need the short fold lines. The long fold lines correspond to the diagonals of the pentagon. Therefore, the long and short fold lines are in the ratio of the golden section, so they are Major and Minor (Fig. 4.15).

However, we can also fit a larger pentagon into the strip.

Accordingly, we can build a large pentagon from our strip (Figs. 4.16 and 4.17). However, with strips that have different colors on both sides, the color change does not work out. This is because five is an odd number.

Our folding strip also provides the solution to the following problem: In an isosceles trapezoid, the short parallel side is as long as the legs, the long parallel side is as long as the diagonals. What shape does this trapezoid have?

The folding strip is composed of exactly such trapezoids (Fig. 4.18). The long parallel side and the diagonals are Majors, the short parallel side and the legs are Minors. The trapezoid is formed from a regular pentagon by cutting off a triangle along a diagonal.

Ten such trapezoids can be assembled into a regular decagon (Fig. 4.19).

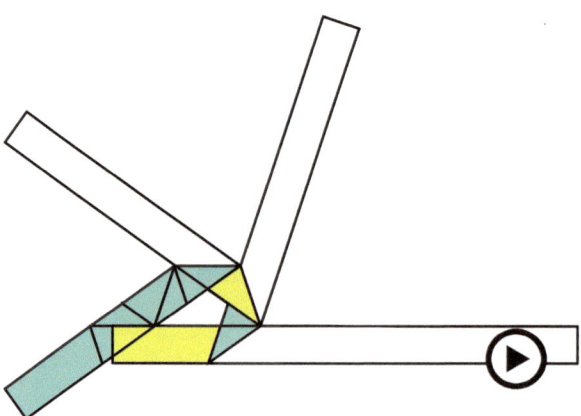

Fig. 4.16 Folding of the large pentagon (▶ https://doi.org/10.1007/000-cb2)

Fig. 4.17   Color change

Fig. 4.18   Trapezoid

Fig. 4.19   Decagon

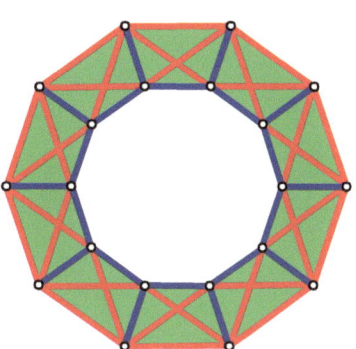

4.1.5   Heptagon

The Golden Ratio also proves itself in our knot example as the simplest non-trivial example of a sequence of further examples. We can, in fact, build one more loop into the knot (Fig. 4.20). This results in a heptagon. It does require some finesse, but it is possible. This is remarkable, because with the usual geometric tools compass and ruler cannot be used to construct the heptagon graphically.—Of course, any graphic software can generate a regular heptagon. However, this is done using computational methods involving trigonometric functions.

With each additional loop, two more corners are created. So, at least theoretically, we can also create the nonagon, hendecagon, tridecagon, etc. as knots. However, the nonagon (Fig. 4.21) is already quite tricky.

Fig. 4.20 Heptagon

Fig. 4.21 Nonagon

4.1.6   Knot with Two Strips

The knot made of two strips (Fig. 4.22) results in a hexagon when tightened. This knot is referred to as True Samaritan Knot or Weaver's Knot or Reef Knot.

A variant is the False Samaritan Knot (Fig. 4.23). In this variant, equilateral triangles of alternating colors are visible in the hexagon.

The two strips for the hexagon knots can be pre-folded to create a sequence of equilateral triangles (Fig. 4.24). To do this, the first fold line is drawn at an angle of 60° and then alternately folded down and up. This method is also a self-correcting algorithm (see [14]).

4.2   The Golden Rectangle

The Golden Rectangle can be produced in various ways by folding.

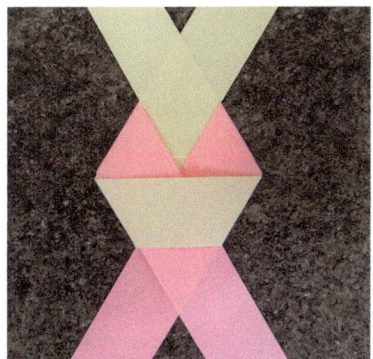

Fig. 4.22   Knot with Two Strips

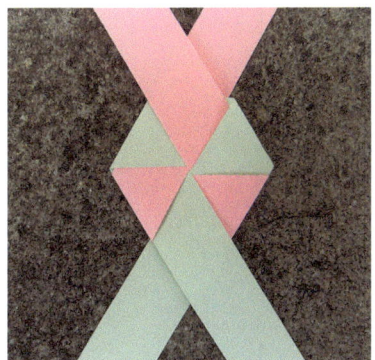

Fig. 4.23   Knot Variant

Fig. 4.24   Equilateral
Triangles

4.2.1   Origami

Origami is the traditional Japanese art of paper folding. From a square sheet of paper, various geometric figures, as well as flowers and animals, are created through folding and occasionally cutting.

This includes the Golden Rectangle. Figure 4.25 shows the essential folding steps for the Golden Rectangle in brief, and Fig. 4.26 shows the complete folding process.

The proof of the consistency of this folding construction is as follows (Fig. 4.27).

Firstly, we have:

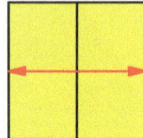 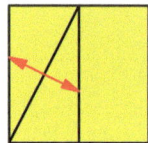 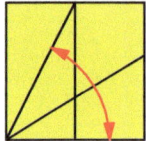 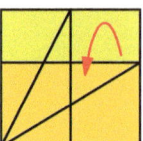

Fig. 4.25 Golden Rectangle

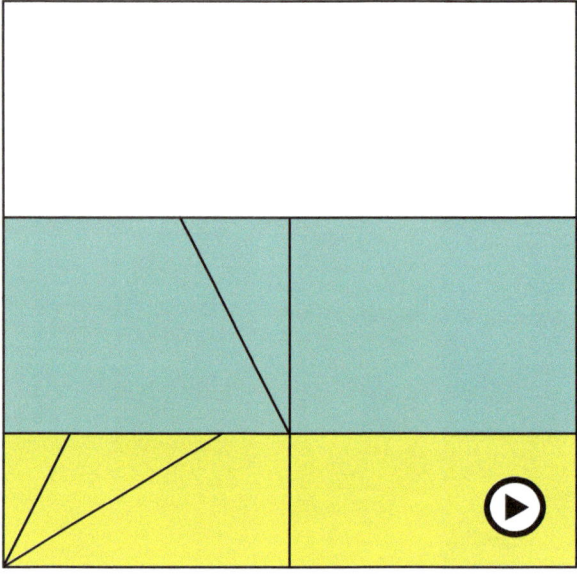

Fig. 4.26 Folding process for the Golden Rectangle (▶ https://doi.org/10.1007/000-cb3)

Fig. 4.27 Proof figure

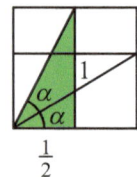 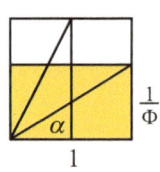

$$\tan(2\alpha) = \frac{1}{\frac{1}{2}} = 2 \tag{4.1}$$

Now we use the addition theorem for the tangent function:

$$\tan(2\alpha) = \frac{2\tan(\alpha)}{1 - \tan(\alpha)^2} \tag{4.2}$$

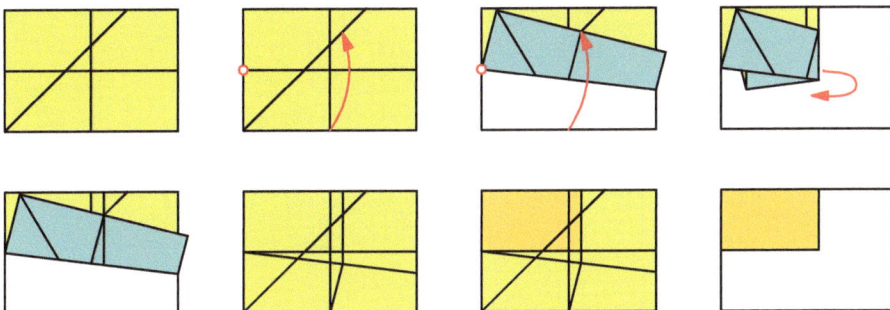

Fig. 4.28 The Golden Rectangle from the DIN format

So we have the condition:

$$\frac{2\tan(\alpha)}{1 - \tan(\alpha)^2} = 2 \tag{4.3}$$

From this, we get the quadratic equation for $\tan(\alpha)$:

$$\tan(\alpha)^2 + \tan(\alpha) - 1 = 0 \tag{4.4}$$

This is the well-known equation for the Golden Ratio. The positive solution is:

$$\tan(\alpha) = \frac{1}{\Phi} \tag{4.5}$$

So we have indeed obtained the Golden Rectangle.

4.2.2 DIN Format

The Golden Rectangle can also be created by folding a piece of paper in DIN format, for example DIN A4. A paper in DIN format has the aspect ratio $\sqrt{2} : 1$.

Figure 4.28 shows the essential steps of folding and unfolding, Fig. 4.29 shows the complete process.

4.3 Regular Pentagon

4.3.1 Exact Folding Construction

We can fold a regular pentagon from a golden rectangle. If we fold the two upper corners into the middle in a golden rectangle in landscape format, angles are created that are exclusively multiples of 9° (Fig. 4.30). In particular, we obtain an isosceles triangle with the angles 54°, 54° and 72°. This is a sector triangle in the regular pentagon.

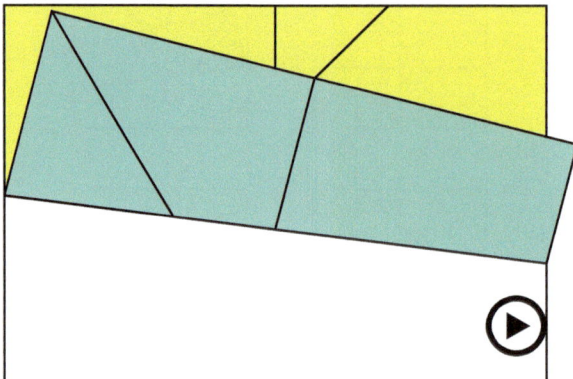

Fig. 4.29 Complete folding process (▶ https://doi.org/10.1007/000-cb1)

Fig. 4.30 Angles in the folded golden rectangle

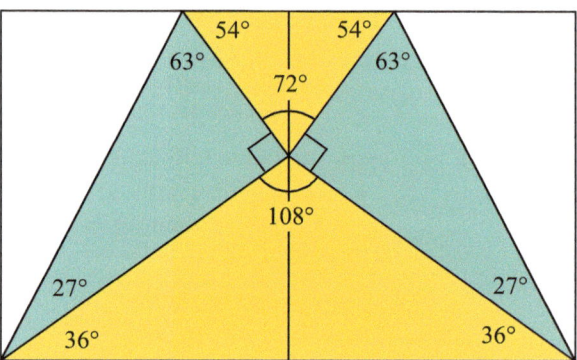

This fact is exploited in the following folding process (Fig. 4.31). A gusset is folded with a sharp angle of 36°, which after unfolding leads to a ten-part fold line bundle with corresponding intermediate angles.

Figure 4.32 shows the entire folding process.

If the 36° gusset is cut before unfolding, a regular pentagon results when unfolded (Figs. 4.33 and 4.34).

If the spickel is figuratively cut on all three edges, we get a silhouette with the symmetries of the regular pentagon (Fig. 4.35).

4.3.2 Approximate Construction

Figure 4.36 shows an approximate construction (student idea) based on a piece of paper in DIN format, for example DIN A4. A paper in DIN format has the aspect ratiol : $\sqrt{2} \approx 1 : 1.414$.

The deviations from the regular pentagon are small (Fig. 4.37). In a regular pentagon, all interior angles measure 108° and all sides are of equal length.

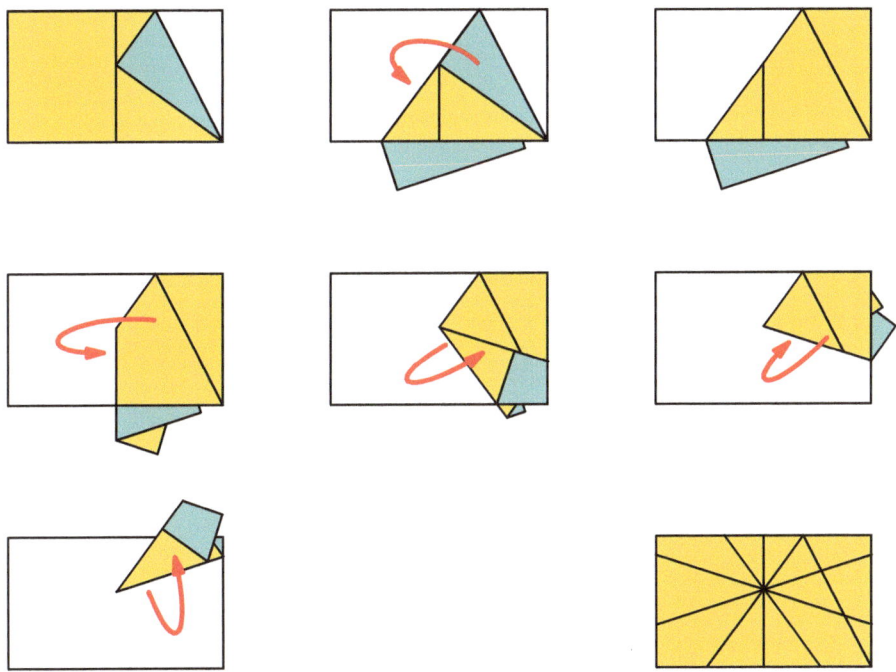

Fig. 4.31   Folding a 36° gusset

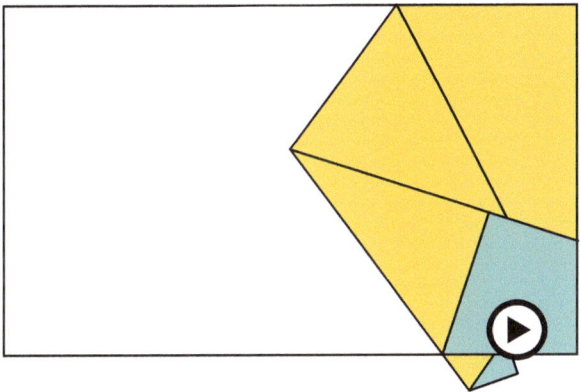

Fig. 4.32   Ten-part fold line bundle (▶ https://doi.org/10.1007/000-cb5)

If we slightly reduce the height of the paper to an aspect ratio 1 : tan (54°) ≈ 1 : 1.376, the same folding process results in an exact pentagon (Fig. 4.38). An angle of 54° can be produced by folding according to Fig. 4.30.

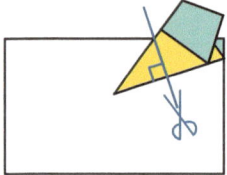  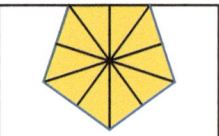

Fig. 4.33 Cutting

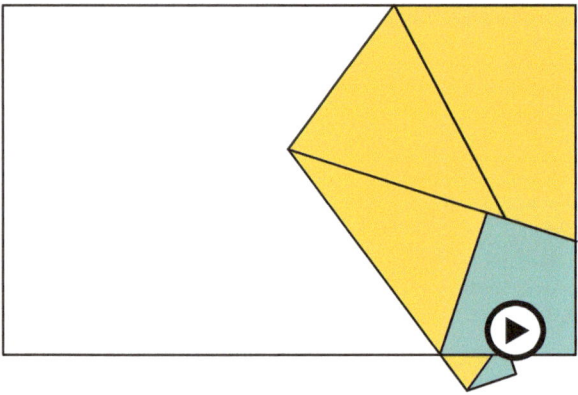

Fig. 4.34 Regular pentagon (▸ https://doi.org/10.1007/000-cb6)

Fig. 4.35 Silhouette

4.4 Star Figures

4.4.1 The Pentagram

From a regular paper pentagon (Figs. 4.33 and 4.34), folding in the corners results in a pentagram (Fig. 4.39).

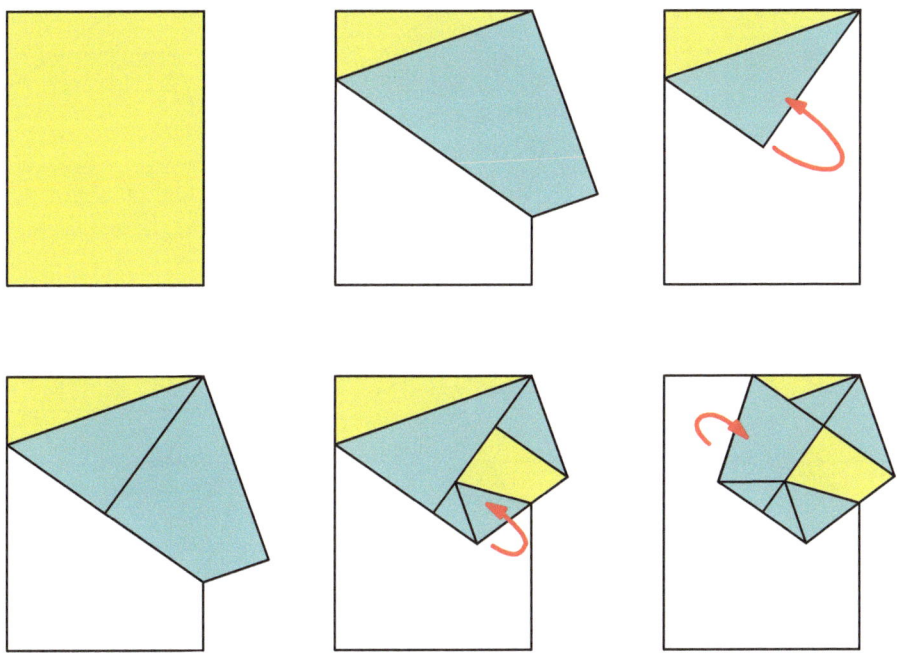

Fig. 4.36 Approximate Construction

Fig. 4.37 Measurements

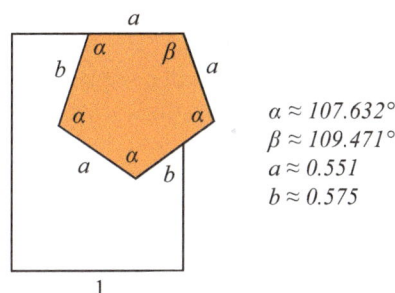

$\alpha \approx 107.632°$
$\beta \approx 109.471°$
$a \approx 0.551$
$b \approx 0.575$

Fig. 4.38 Exact Pentagon

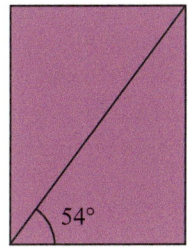

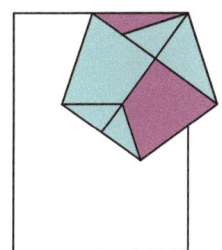

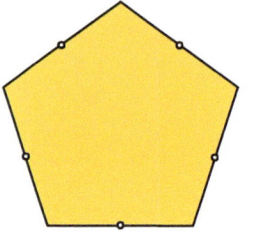 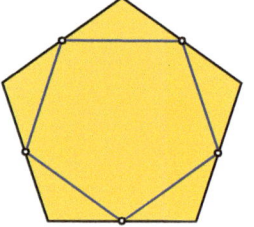 

Fig. 4.39 Pentagram as a folding figure

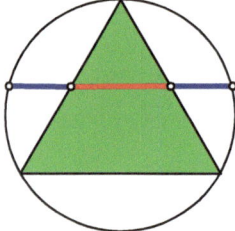 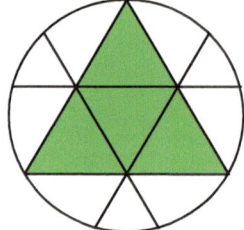 

Fig. 4.40 Figure of Odom

4.4.2   The Figure of Odom

The construction of the Golden Section according to George Odom (1941–2010) is based on an equilateral triangle and its circumcircle (Fig. 4.40). A straight line through two midpoints of the sides, together with the intersection points with the circumcircle, leads to a division in the Golden Section in the order Minor-Major-Minor. We can carry out this construction with all midpoints of the sides and connect the points on the circumcircle with further chords.

This is the starting figure for the following crafting suggestion: We divide an equilateral triangle in height according to the Golden Section with the Major on top and fold the base trapezoid to the back (Fig. 4.41). This creates a component.

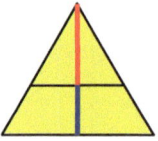 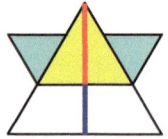 

Fig. 4.41 Component and Star

We need three equally sized components and assemble them along the fold lines so that the three fold lines themselves form an equilateral triangle.

This creates a star with new tips, all of which lie on a circular line; that's the trick of the matter.

Number Sequences

5

5.1　Linearization of Powers of the Golden Ratio

Since Φ is a solution of the quadratic equation

$$x^2 = x + 1 \tag{5.1}$$

the following applies:

$$\Phi^2 = \Phi + 1 \tag{5.2}$$

Figure 5.1 provides a visualization of this fact. We see on the one hand a square with side length Φ, thus the area Φ^2, and on the other hand a Golden Rectangle with the sides Φ and 1, thus the area Φ, as well as a square with side length 1 and thus the area 1. Both figures can be broken down into the same area pieces.

So, we can replace Φ^2 with a linear expression in Φ. This also allows us to replace higher powers higher powers of Φ with a linear expression in Φ. For example:

$$\Phi^3 = \Phi^2\Phi = (\Phi + 1)\Phi = \Phi^2 + \Phi = (\Phi + 1) + \Phi = 2\Phi + 1 \tag{5.3}$$

The third power Φ^3 can, however, be calculated more simply by the following consideration. From

$$\Phi^2 = \Phi + 1 \tag{5.4}$$

by multiplying with Φ we initially get:

$$\Phi^3 = \Phi^2 + \Phi \tag{5.5}$$

Now, we just need to replace the right side for Φ^2 with the linear expression in Φ.

Supplementary Information The electronic version of this chapter contains additional material, which can be accessed via the following link https://doi.org/10.1007/978-3-662-69890-7-0_5. The videos can be played by clicking on the DOI link in the legend of a corresponding figure, or by scanning this link with the SN More Media App.

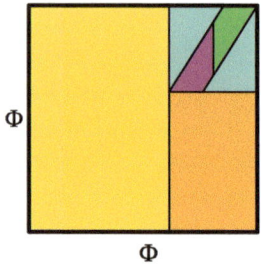 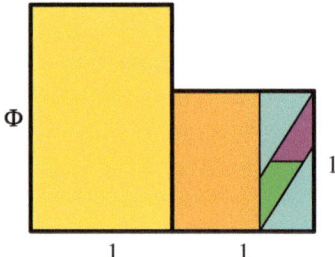

Fig. 5.1 Common Decomposition

In general, from

$$\Phi^2 = \Phi + 1 \tag{5.6}$$

by multiplying with Φ^n we get the relationship:

$$\Phi^{n+2} = \Phi^{n+1} + \Phi^n \tag{5.7}$$

If the linear expressions for Φ^{n+1} and Φ^n are known, the linear expression for Φ^{n+2} results from their addition. Explicitly, the following applies:

$$
\begin{aligned}
\Phi^0 &= & 1 &= 1 \\
\Phi^1 &= & \Phi &= \Phi \\
\Phi^2 &= & \Phi + 1 &= \Phi + 1 \\
\Phi^3 &= \Phi^2 + \Phi &= 2\Phi + 1 \\
\Phi^4 &= \Phi^3 + \Phi^2 &= 3\Phi + 2 \\
\Phi^5 &= \Phi^4 + \Phi^3 &= 5\Phi + 3 \\
\Phi^6 &= \Phi^5 + \Phi^4 &= 8\Phi + 5
\end{aligned}
\tag{5.8}
$$

The new row is always the sum of the two preceding rows. This linearization of the powers of Φ was already known to Leonhard Euler(1707–1783).

The coefficients f_n in

$$\Phi^n = f_n \Phi + f_{n-1}, n \in \{2, 3, 4, \dots\} \tag{5.9}$$

are the so-called Fibonacci numbers. For the Fibonacci numbers, the following recursion obviously applies

$$f_{n+2} = f_{n+1} + f_n \tag{5.10}$$

with the starting values $f_1 = 1$ and $f_2 = 1$.

An analogous consideration yields the linearization formula for the powers of the number $\left(-\frac{1}{\Phi}\right)$, which satisfies the condition

$$\left(-\frac{1}{\Phi}\right)^2 = \left(-\frac{1}{\Phi}\right) + 1 \tag{5.11}$$

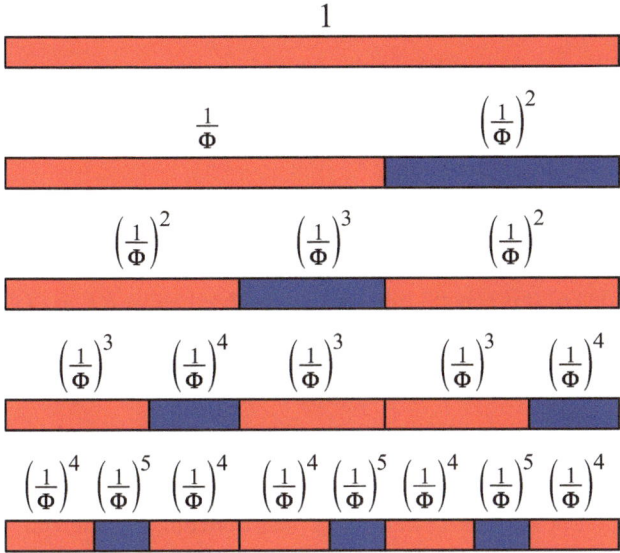

Fig. 5.2 Continued division of the unit distance

$$\vdots$$

$$\left(-\frac{1}{\Phi}\right)^n = f_n\left(-\frac{1}{\Phi}\right) + f_{n-1}, n \in \{2,3,4,\dots\} \qquad (5.12)$$

A distance of length 1 can be divided using the powers of $\left(\frac{1}{\Phi}\right)$ according to Fig. 5.2 divide. From the minors, in each subsequent division step, the majors are formed. That's how it can go.

Figure 5.3 focuses on the visual structure of this division.

Fig. 5.3 Visual structure

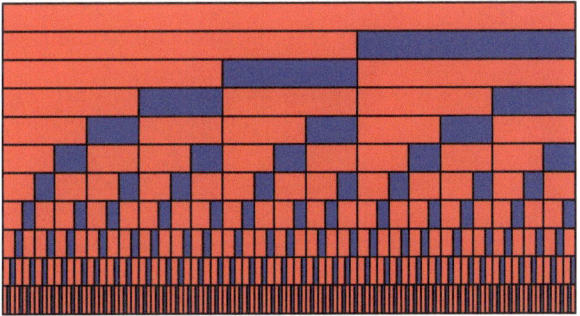

5.2 Fibonacci Sequences

We understand the (special) Fibonacci sequence to be the sequence with the recursive formation law

$$f_{n+2} = f_{n+1} + f_n \tag{5.13}$$

and the starting values $f_1 = 1$ and $f_2 = 1$, so the sequence:

$$1, 1, 2, 3, 5, 8, 13, 21, 34, 55, \ldots \tag{5.14}$$

Generalizations of this special Fibonacci sequence can be found in [7, 20, 21, 27].

The name "Fibonacci" is a shortening of "Filius Bonacci", which means "son of Bonacci". His actual name was Leonardo of Pisa, he was born between 1170 and 1180, and learned everything known about calculation methods at the time on trading trips, often accompanied by his father, which took him to Algeria, Egypt, Syria, Greece, Sicily, and Provence. His epochal work Liber Abaci, published in 1202, introduced Indian arithmetic to Europe and introduced the now common Arabic notation of numbers. Fibonacci's year of death is unknown.

The numbers of the Fibonacci sequence appeared in the linearization of the powers of Φ. To obtain an explicit formula for the numbers of this Fibonacci sequence, we form the difference of the linearization formulas Eqns. 5.9 and 5.12 of Φ^n and $\left(-\frac{1}{\Phi}\right)^n$. This results in

$$\Phi^n - \left(-\frac{1}{\Phi}\right)^n = f_n \left(\Phi - \frac{1}{\Phi}\right) \tag{5.15}$$

and due to $\left(\Phi - \frac{1}{\Phi}\right) = \sqrt{5}$ the formula:

$$f_n = \frac{1}{\sqrt{5}} \left(\Phi^n - \left(-\frac{1}{\Phi}\right)^n\right) \tag{5.16}$$

This formula is named after Jacques Philippe Marie Binet (1786–1856), but was already known to (1700–1782).

The Fibonacci sequence $\{1, 1, 2, 3, 5, 8, 13, \ldots\}$ is thus the difference of two geometric sequences with the quotients Φ and $\left(-\frac{1}{\Phi}\right)$. Due to $\Phi > 1$ and $\left|-\frac{1}{\Phi}\right| < 1$ tending towards large indices n the proportion $\left(-\frac{1}{\Phi}\right)^n$ against 0. It then applies:

$$f_n \approx \frac{1}{\sqrt{5}} \Phi^n \tag{5.17}$$

The numbers of the Fibonacci sequence can thus be approximated by powers of the golden ratio. For large indices n, the Fibonacci sequence behaves like a geometric sequence with the quotient Φ.

For the quotient of two consecutive Fibonacci numbers, we therefore obtain the limit:

Tab. 5.1 Quotients

n	1	2	3	4	5	6
f_n	1	1	2	3	5	8
$\frac{f_{n+1}}{f_n}$	$\frac{1}{1} = 1$	$\frac{2}{1} = 2$	$\frac{3}{2} = 1.5$	$\frac{5}{3} = 1.\overline{6}$	$\frac{8}{5} = 1.6$	$\frac{13}{8} = 1.625$

Quotients

Tab. 5.2 Backward Calculation

z	−7	−6	−5	−4	−3	−2	−1	0	1	2	3	4	5	6	7
f_z	13	−8	5	−3	2	−1	1	0	1	1	2	3	5	8	13

Backward calculation

$$\lim_{n \to \infty} \frac{f_{n+1}}{f_n} = \Phi \tag{5.18}$$

Thus, the golden ratio can be approximated by the quotient of two consecutive Fibonacci numbers, as illustrated in Tab. 5.1.

The Fibonacci sequence can also be calculated "backwards" from the starting values (Tab. 5.2).

It is $f_0 = 0$ and $f_{-n} = (-1)^{n+1} f_n$. The Binet formula also applies to negative indices.

5.2.1 Golden Trapezoid and Golden Star

The recursion formula $f_{n+2} = f_{n+1} + f_n$ of the Fibonacci sequence can be visualized on an isosceles trapezoidwith base angle 60° (Fig. 5.4). For this, we choose the top parallel f_n and the two legs f_{n+1}. Then the base parallel has the length f_{n+2}, as we can see by drawing an equilateral triangle (cf. [37, 39]).

Figure 5.5 shows the first Fibonacci trapezoids. Because of $f_0 = 0$, the foremost example is just a triangle (top parallel is zero).

The Fibonacci trapezoids can be triangulated with equilateral triangles. The numbers of equilateral triangles needed for this triangulation are in order: $1, 3, 8, 21, 55, \ldots$. We recognize a subsequenceof the Fibonacci sequence.

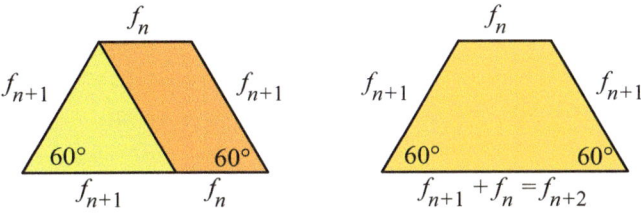

Fig. 5.4 Fibonacci Trapezoid

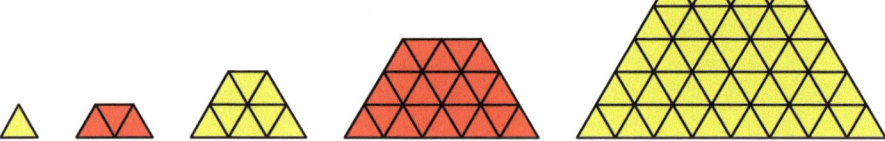

Fig. 5.5 Fibonacci Trapezoids

Six sets of Fibonacci trapezoids can be assembled into a star, which fits into a triangle grid (Fig. 5.6).

We now normalize the leg length of the Fibonacci trapezoids to 1, by dividing all side lengths by f_{n+1}, and then make the limit transition $n \to \infty$. This results in the Golden Trapezoid. It has the top parallel $\frac{1}{\Phi}$ and the base parallel Φ (Fig. 5.7).

From the measurements of the Golden Trapezoid (Fig. 5.8), the length of the diagonal d results:

$$d^2 = \left(\frac{\sqrt{3}}{2}\right)^2 + \left(\frac{1}{\Phi} + \frac{1}{2}\right)^2 = \frac{3}{4} + \underbrace{\left(\frac{1}{\Phi}\right)^2 + \frac{1}{\Phi}} + \frac{1}{4} = 2 \qquad (5.19)$$

Thus, $d = \sqrt{2}$. In the Golden Trapezoid, the numbers 1, $\sqrt{2}$ (diagonal), $\sqrt{3}$ (at the trapezoid height) and $\sqrt{5}$ (Golden Ratio) appear.

Corresponding to the Fibonacci star, the Golden Star can be composed of golden trapezoids, the lengths of which form a geometric sequence with the factor

Fig. 5.6 Fibonacci Star

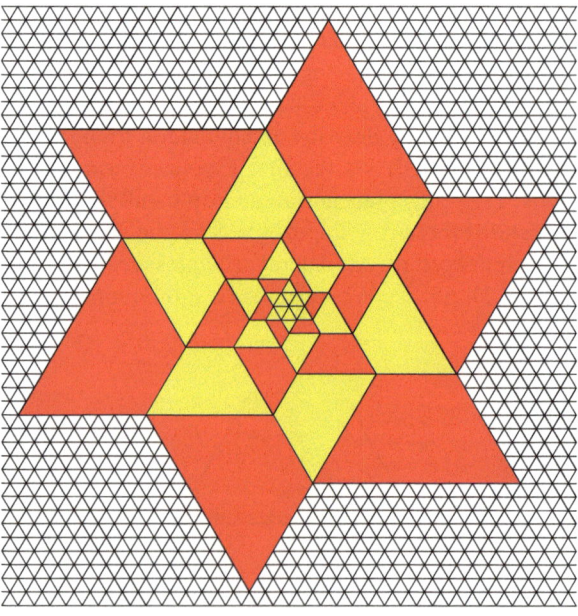

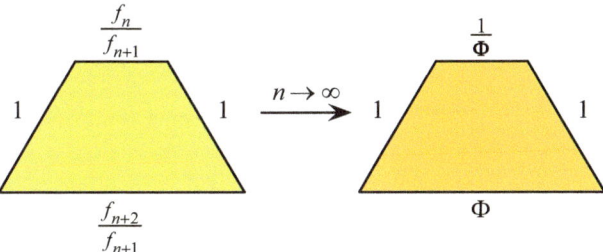

Fig. 5.7 Golden Trapezoid

Fig. 5.8 Lengths in the
Golden Trapezoid

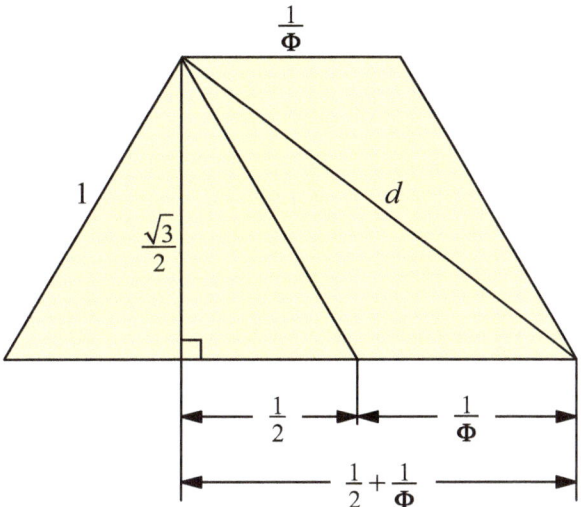

Φ (Fig. 5.9). However, due to the irrationality of Φ, it does not fit into a triangular grid.

5.2.2   Family Tree of a Drone

The family treeof a drone provides an illustration of the Fibonacci sequence. Since a drone comes from an unfertilized bee egg, and a queen or a worker beecomes from a fertilized egg (the latter depends on the diet), a drone has only one maternal parent, while a queen has two parents (Fig. 5.10).

From this, the family tree of a drone is shown in Fig. 5.11. This family tree is asymmetric, with the number of female ancestors predominating.

For the n-th parent generation, there are f_n females and f_{n-1} drones, the female proportion tends towards $n \to \infty$ for $\frac{1}{\Phi}$.

Fig. 5.9 Golden Star

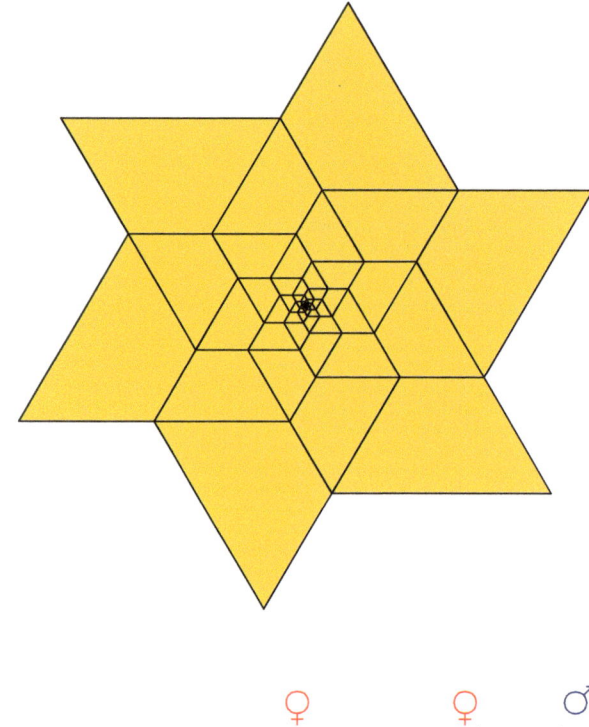

Fig. 5.10 Parents of a drone
and a queen bee

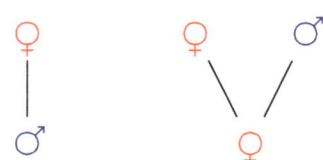

5.2.3 Approximation of the Golden Rectangle by Fibonacci Squares

In the subdivision of the Golden Rectangle into squares, due to the irrationality of the Golden Ratio, there can be no smallest square where the process stops.

We now consider the reverse, what happens if we start with a real smallest square, to which we assign a side length of one, and build up the figure into a rectangle by attaching further squares. Figure 5.12 shows the first five steps of this successive attachment of squares.

The successive squares have side lengths of 1, 1, 2, 3, 5, 8, ... Each new square side is, as can also be seen from Fig. 5.12, the sum of the sides of the two preceding squares. The sequence of side lengths is therefore the Fibonacci sequence. The resulting rectangles have two consecutive Fibonacci numbers as side lengths. Since the ratio of two consecutive Fibonacci numbers tends towards the Golden Ratio, the rectangles constructed in this way are approximations of the Golden Rectangle.

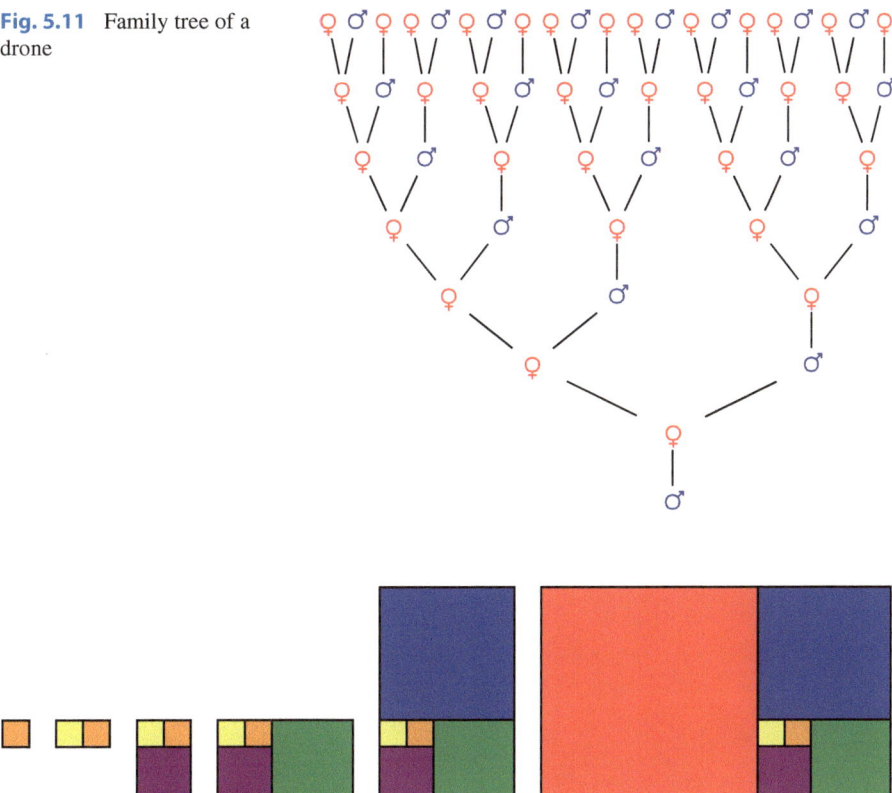

Fig. 5.11   Family tree of a drone

Fig. 5.12   Attachment of squares

The application of the Euclidean algorithm to these rectangles leads back to the smallest square with a side length of one. Two consecutive Fibonacci numbers therefore only have the number one as a common divisor; they are coprime. Other Fibonacci numbers can indeed have common divisors. For example, every third Fibonacci number is an even number. These numbers therefore all have the common divisor 2. Every fourth Fibonacci number is divisible by 3. Every fifth Fibonacci number is divisible by 5. In general, f_n divides the number f_m, if m is an integer multiple of n.

Figure 5.13 shows a corner-to-corner arrangement of the Fibonacci squares.

5.2.4   In Pascal's Triangle

Fig. 5.14 shows Pascal's triangle of binomial coefficients.

In the diagonal slopes, there are only ones. Furthermore, each number is the sum of the two numbers to the right and left above. So, for example, the fifteen in

Fig. 5.13 Corner-to-corner arrangement

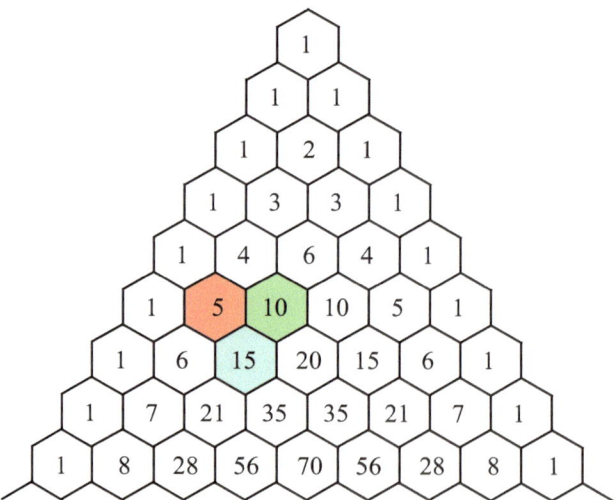

Fig. 5.14 Binomial coefficients

the blue field is the sum of the ten in the green field and the five in the red field. This calculation method also applies at the edge of the triangle, if we assume that we only have zeros outside the triangle.

And now we add up those numbers that are in the same diagonal (Fig. 5.15). The diagonals have an inclination angle of 30° compared to the horizontal. Three diagonals are highlighted in color. The sum of the numbers in the red diagonal is

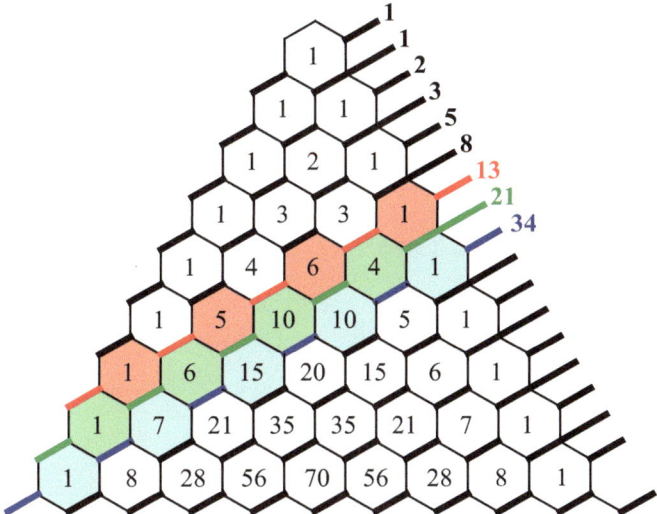

Fig. 5.15 Diagonal sums

13, in the green diagonal we have the sum 21 and in the blue diagonal the sum 34. However, it is the case that each number in the blue diagonal is the sum of the numbers above it in the green and red diagonal. Each number is used exactly once in both the green and red diagonal for the blue diagonal. Therefore, the sum 34 must also be composed of the sums 21 and 13 of the green and red diagonal, so $34 = 21 + 13$. But this is exactly the recursion formula for the Fibonacci numbers. Since we have two ones at the top of the triangle, which serve as starting values, the diagonal sums are generally the Fibonacci numbers.

5.2.5   Arbitrary Starting Values

Let's start a sequence with the Fibonacci recursion formula

$$b_{n+2} = b_{n+1} + b_n \tag{5.20}$$

with arbitrary starting values $b_1 = d$ and $b_2 = c$, the result is:

$$
\begin{aligned}
b_1 &= & d \\
b_2 &= & c \\
b_3 &= & c + d \\
b_4 &= & 2c + d \\
b_5 &= & 3c + 2d \\
b_6 &= & 5c + 3d \\
b_7 &= & 8c + 5d
\end{aligned}
\tag{5.21}
$$

It is obvious:

$$b_n = f_{n+1}c + f_{n-2}d \tag{5.22}$$

Where f_n is the special Fibonacci sequence with the starting values $f_1 = 1$ and $f_2 = 1$.

From Binet's formula, the explicit formula for $n > 1$ follows:

$$b_n = \frac{1}{\sqrt{5}}\left((c\Phi + d)\Phi^{n-2} - \left(-c\frac{1}{\Phi} + d\right)\left(-\frac{1}{\Phi}\right)^{n-2} \right) \tag{5.23}$$

In the case of $c\Phi + d \neq 0$ for large n:

$$b_n \approx \frac{c\Phi + d}{\sqrt{5}}\Phi^{n-2} \tag{5.24}$$

We again obtain the limit:

$$\lim_{n\to\infty} \frac{b_{n+1}}{b_n} = \Phi \tag{5.25}$$

The quotient sequenceof a sequence with the Fibonacci recursion formula generally has the golden ratio as its limit. This limit is independent of the starting values. Only the recursion formula is relevant for the limit.

One might now ask whether there is a Fibonacci sequence that is also a geometric sequence. We insert the approach

$$b_n = aq^n \tag{5.26}$$

into the recursion formula:

$$aq^{n+2} = aq^{n+1} + aq^n \tag{5.27}$$

This results in:

$$q^2 = q + 1 \tag{5.28}$$

The quotient q of the sequence must therefore be $q = \Phi$ or $q = -\frac{1}{\Phi}$.

5.3   Continued Fractions

Fig. 5.16 shows five similar right-angled triangles. Four of them (three yellow and one in magenta) are even congruent. The fifth triangle (light blue) deviates. It is usually smaller or larger.

When exactly is the light blue triangle the same size, i.e., congruent to the others?

To understand this, we fit a coordinate system into the outline of the general figure (Fig. 5.17). The circle is the unit circle. The two relevant points on the x-axis have the x-coordinates p and q.

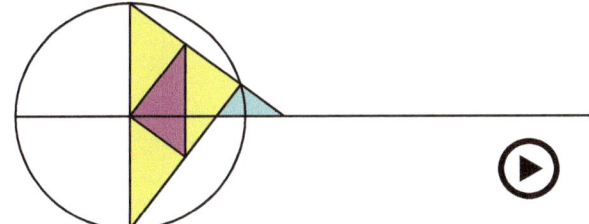

Fig. 5.16 Kinematics (▶ https://doi.org/10.1007/000-cb7)

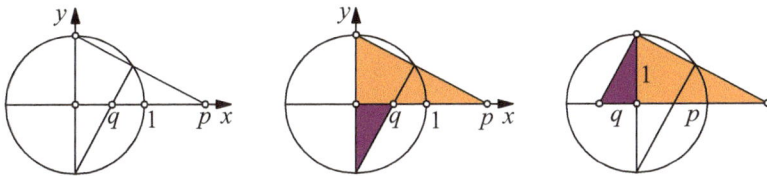

Fig. 5.17 Reflection on the unit circle

Now we color two new right-angled triangles (orange and purple). The orange triangle has the long leg p and the purple triangle has the short leg q.

Finally, we reflect the purple triangle at the center of the circle.

The two triangles now form a large right-angled triangle together. It has the hypotenuse segments p and q and the height 1.

According to the altitude theorem, $pq = 1$. Thus, p and q are reciprocals of each other, so:

$$q = \frac{1}{p} \tag{5.29}$$

This construction is referred to as reflection on the unit circle.

The four already congruent triangles have the hypotenuse length 1. This must now also apply to the light blue triangle in our problem, so:

$$p - q = 1 \tag{5.30}$$

Because of $q = \frac{1}{p}$ we get:

$$p - \frac{1}{p} = 1 \tag{5.31}$$

From this we obtain the quadratic equation:

$$p^2 - p - 1 = 0 \tag{5.32}$$

The solution relevant to us is $p = \Phi$, that is, the Golden Ratio. Therefore, we can appropriately draw Major and Minor into the solution figure (Fig. 5.18).

We can also write the condition for the congruence of the light blue triangle in the following form:

Fig. 5.18 Five congruent triangles

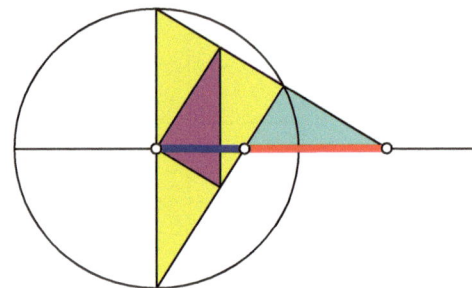

$$p = 1 + \frac{1}{p} \tag{5.33}$$

From this, we construct a recursion formula:

$$c_{n+1} = 1 + \frac{1}{c_n} \tag{5.34}$$

With the starting value $c_1=1$ we obtain in sequence:

$$
\begin{aligned}
c_1 &= & 1 \\
c_2 &= & 1 + \tfrac{1}{1} = 2 \\
c_3 &= & 1 + \cfrac{1}{1+\frac{1}{1}} = \tfrac{3}{2} = 1.5 \\
c_4 &= & 1 + \cfrac{1}{1+\cfrac{1}{1+\frac{1}{1}}} = \tfrac{5}{3} = 1.\bar{6} \\
c_5 &= & 1 + \cfrac{1}{1+\cfrac{1}{1+\cfrac{1}{1+\frac{1}{1}}}} = \tfrac{8}{5} = 1.6 \\
c_6 &= 1 + & \cfrac{1}{1+\cfrac{1}{1+\cfrac{1}{1+\frac{1}{1}}}} = \tfrac{13}{8} = 1.625
\end{aligned}
\tag{5.35}
$$

We recognize the quotient sequence of the Fibonacci numbers. This tends towards the Golden Ratio Φ. Therefore, the Golden Ratio can also be represented as a so-called continued fraction:

$$\Phi = 1 + \cfrac{1}{1 + \cfrac{1}{1+\cfrac{1}{1+\frac{1}{1+\cdots}}}} \tag{5.36}$$

The representation contains only ones. With other numbers, other limit values result.

For example:

$$2 + \cfrac{1}{2 + \cfrac{1}{2+\cfrac{1}{2+\frac{1}{2+\cdots}}}} = 1 + \sqrt{2} \tag{5.37}$$

We can verify this by solving the equation $x = 2 + \frac{1}{x}$.

On the other hand (swapping the numbers 1 and 2) is:

$$1 + \cfrac{2}{1 + \cfrac{2}{1 + \cfrac{2}{1 + \cfrac{2}{1 + \cdots}}}} = 2 \tag{5.38}$$

This is the relevant solution of the equation $x = 1 + \frac{2}{x}$. It is not always easy to count to two.

5.4   Chain Roots

Analogous to the continued fractions, we can proceed with the chain roots. We work with the recursion formula:

$$w_{n+1} = \sqrt{1 + w_n} \tag{5.39}$$

With the initial value $w_1 = 1$ we get telescopicTelescope formulas:

$$\begin{aligned}
w_1 &= 1 \\
w_2 &= \sqrt{1 + 1} \approx 1.414213562 \\
w_3 &= \sqrt{1 + \sqrt{1 + 1}} \approx 1.553773974 \\
w_4 &= \sqrt{1 + \sqrt{1 + \sqrt{1 + 1}}} \approx 1.598053182 \\
w_5 &= \sqrt{1 + \sqrt{1 + \sqrt{1 + \sqrt{1 + 1}}}} \approx 1.611847754 \\
w_6 &= \sqrt{1 + \sqrt{1 + \sqrt{1 + \sqrt{1 + \sqrt{1 + 1}}}}} \approx 1.616121206
\end{aligned} \tag{5.40}$$

We suspect:

$$w = \lim_{n \to \infty} w_n = \Phi \tag{5.41}$$

To see this, we insert the hypothetical limit value w into the recursion formula and obtain the root equation:

$$w = \sqrt{1 + w} \tag{5.42}$$

By squaring, we get the quadratic equation:

$$w^2 = 1 + w \tag{5.43}$$

The solution relevant to us is $w = \Phi$.

Therefore, we can write the golden ratio as a chain root:

$$\Phi = \sqrt{1 + \sqrt{1 + \sqrt{1 + \sqrt{1 + \cdots}}}} \tag{5.44}$$

In this chain root, we only have ones. We can change the numbers.
For example:

$$\sqrt{6+\sqrt{6+\sqrt{6+\sqrt{6+\cdots}}}} = 3 \tag{5.45}$$

This can be checked with the equation $x = \sqrt{6+x}$.
We can also incorporate coefficients in the roots:

$$\sqrt{5+4\sqrt{5+4\sqrt{5+4\sqrt{5+\cdots}}}} = 5 \tag{5.46}$$

The control equation is $x = \sqrt{5+4x}$.

Regular and Related Solids

6

6.1 The Regular Solids

The cube is bounded by six congruent squares, at each corner of the cube three of these side squares meet. Generally, we speak of a regular solid when it is convex, i.e., it contains no indentations, is bounded by congruent equilateral polygons (side regularity) and when the same number of side faces meet at all corners (corner regularity). These requirements are very restrictive, as there are only five regular solids (Fig. 6.1), namely the tetrahedron bounded by four equilateral triangles, the octahedron bounded by eight equilateral triangles bounded octahedron, the cube, which is bounded by six squares, the icosahedron bounded by 20 equilateral triangles bounded icosahedron and finally the dodecahedron bounded by 12 regular pentagons bounded dodecahedron. These five regular solids, also known as Platonic solids, have played an important role since antiquity not only in geometry, but also in natural sciences, which deal with spatial structures, such as crystallography or the description of molecular arrangements.

The dodecahedron and the icosahedron contain regular pentagons. In the dodecahedron, these are the side faces. In the icosahedron, five triangles meet at one corner; the sides of these triangles that do not run into this corner form a regular pentagon. Thus, the golden ratio will also occur in these two regular solids. For example, if the icosahedron and dodecahedron each stand on one side face, the level heights of the vertices are in the ratio of the golden section (Fig. 6.2).

Supplementary Information The online version contains supplementary material available at https://doi.org/10.1007/978-3-662-69890-7_6.

Fig. 6.1 The five regular solids

Fig. 6.2 Level heights

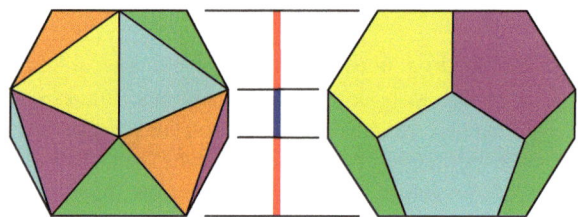

6.2   Constructions Based on the Cube and Octahedron

The icosahedron or the dodecahedron can be inscribed or circumscribed to the cube and the octahedron in a suitable manner.

6.2.1   In the Cube

As an example, an icosahedron with edge length 2 should be inscribed in a cube according to Fig. 6.3.

We denote the side length of the icosahedron with 2 s. Thus, we obtain for the marked vertices A, B and C the coordinates $A(1, -s, 0)$, $B(1, s, 0)$ and $C(s, 0, 1)$.

The triangle ABC should be equilateral with the side length $2s$; the distance BC must have the length $2s$, thus:

$$4s^2 = (s - 1)^2 + s^2 + 1 \tag{6.1}$$

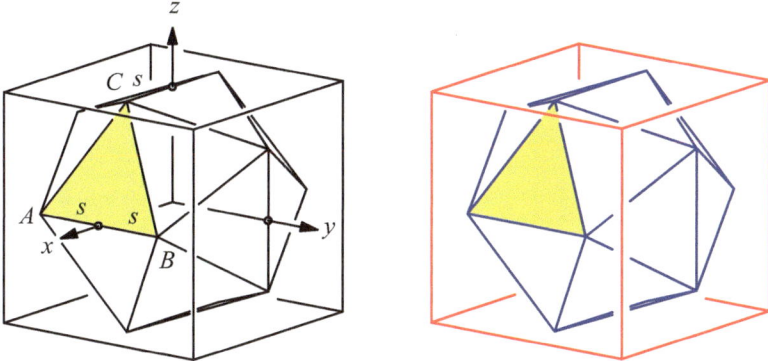

Fig. 6.3   Icosahedron in the cube

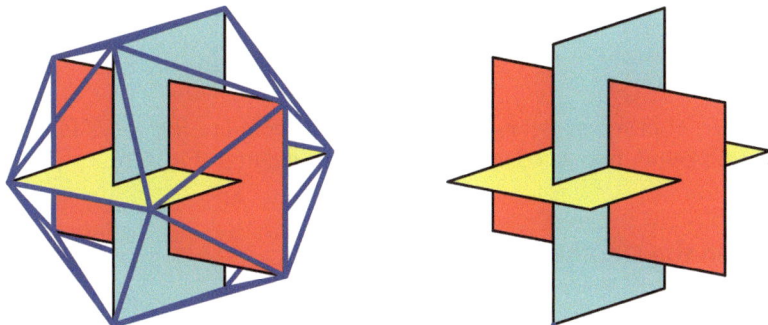

Fig. 6.4   Golden Rectangles in the Icosahedron

This condition can be transformed to:

$$s^2 + s - 1 = 0 \tag{6.2}$$

This equation has the two solutions $s_1 = \frac{1}{\Phi}$ and $s_2 = -\Phi$.

The positive first solution corresponds to the icosahedron the icosahedron of Fig. 6.3. The edge lengths of the cube and the inscribed icosahedron are in the ratio of Major to Minor.

This implies that the intersecting rectangles marked in Fig. 6.4 have the aspect ratio of the Golden Ratio, thus they are Golden Rectangles.

Such a framework of an icosahedron, consisting of three Golden Rectangles framework, can easily be made from three Golden cardboard rectangles with a longitudinal slot in the middle; for assembly reasons, however, one of the three slots must be led to the edge (Fig. 6.5).

The framework can be supplemented to an icosahedron by threads fixed with pins at the corners of the rectangle.

Fig. 6.5 Components for the Icosahedron Framework

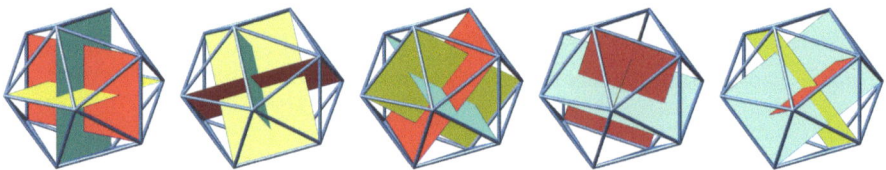

Fig. 6.6 Five Frameworks in the Icosahedron

Apart from different arrangements of colors, there are five ways to incorporate such a framework into an edge model of an icosahedron (Fig. 6.6).

6.2.2   In the Octahedron

We divide the twelve edges of a regular octahedron each in the ratio of the Golden Section (Fig. 6.7). Two majors and two minors should alternate at each corner of the octahedron. The twelve dividing points are now the corners of an icosahedron ([4], p. 52).

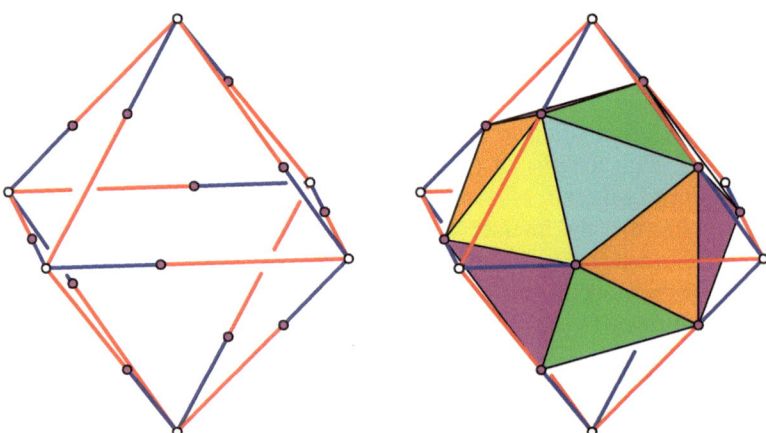

Fig. 6.7 Icosahedron in the Octahedron

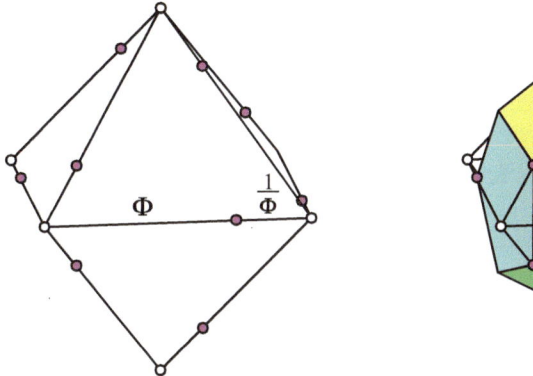

Fig. 6.8 Dodecahedron in the Octahedron

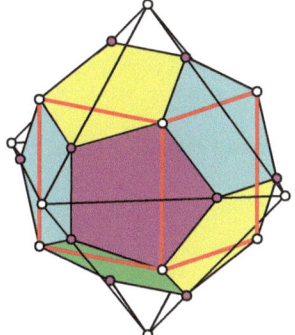

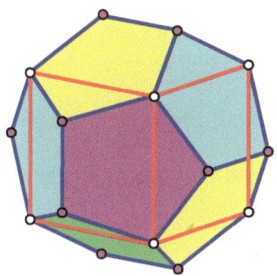

Fig. 6.9 Cube in the Dodecahedron

But we can also divide the edges of an octahedron in the ratio $\Phi : \frac{1}{\Phi}$ (Fig. 6.8). This is no longer the Golden Section, but, if you like, the High Golden Section. This ratio can be written in various ways:

$$\Phi : \frac{1}{\Phi} = \Phi^2 : 1 = (1 + \Phi) : 1 \approx 2.618 : 1 \qquad (6.3)$$

Now a regular dodecahedron can be fitted in such a way that twelve of the 20 corners of the dodecahedron fall on the dividing points of the twelve octahedron edges. The remaining eight corners of the dodecahedron lie outside the octahedron. However, they can be connected to a cube (Fig. 6.9). Although the corners of the cube are outside the octahedron, the edges of the cube run within the edges of the octahedron. The cube edges relate to the dodecahedron edges as Major to Minor.

6.3   Diagonals

The diagonals in icosahedra and dodecahedra lead to star figures.

6.3.1   Diagonals in the Icosahedron

The center and the twelve vertices of an icosahedron can be described in a Cartesian coordinate system with the following coordinates:

$$A_0 = (0,0,0) \tag{6.4}$$

$$A_1 = (\Phi, 1, 0) \quad A_2 = (\Phi, -1, 0) \quad A_3 = (-\Phi, 1, 0) \quad A_4 = (-\Phi, -1, 0)$$
$$A_5 = (1, 0, \Phi) \quad A_6 = (1, 0, -\Phi) \quad A_7 = (-1, 0, \Phi) \quad A_8 = (-1, 0, -\Phi)$$
$$A_9 = (0, \Phi, 1) \quad A_{10} = (0, \Phi, -1) \quad A_{11} = (0, -\Phi, 1) \quad A_{12} = (0, -\Phi, -1)$$

Figure 6.10 shows the position of the points in the coordinate system with the corner numbers. The icosahedron stands on the edge $A_6 A_8$ (an unstable equilibrium).

Figure 6.11 shows the icosahedron with all diagonals. Since we have twelve vertices, there are $\binom{12}{2} = 66$ connections between these vertices. However, the 30 icosahedron edges are included.

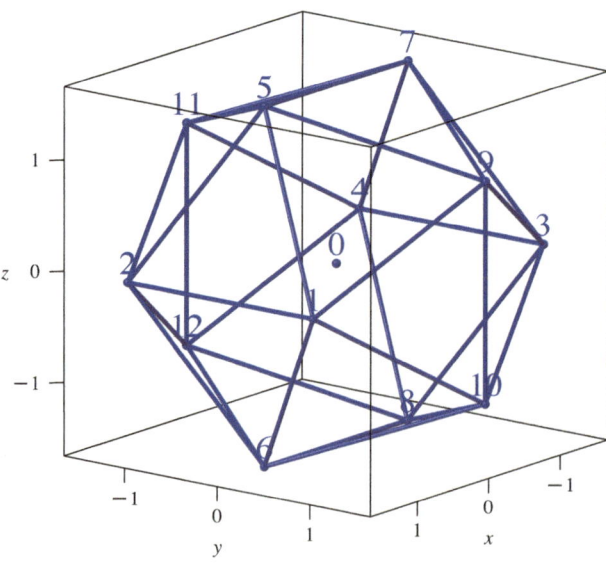

Fig. 6.10   Corner numbers in the icosahedron

Fig. 6.11 Edges and
diagonals in the icosahedron

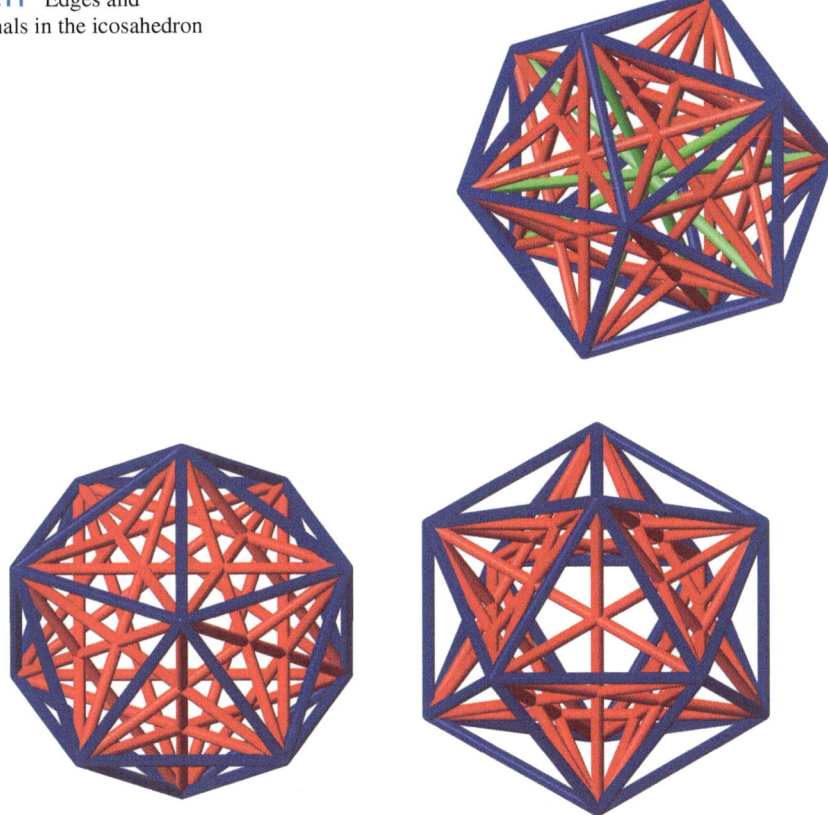

Fig. 6.12 Special views

Tab. 6.1 Edges and diagonals in the icosahedron

Farbe	Anzahl	Länge exakt	Länge numerisch	
	30	2	2	Kanten
	30	2Φ	3,236067977	Raumdiagonale
	6	$2\sqrt{2+\Phi}$	3,804226065	Mittelpunktdiagonale

Edges and Diagonals

Figure 6.12 shows special views of the icosahedron and its diagonals. In both examples, the center diagonals are hidden.

Table 6.1 contains an overview of the edges and diagonals.

Now we fill the spaces between the edges and diagonals with geometric surfaces, that is, with triangles, pentagons, and pentagrams.

Fig. 6.13 Equilateral triangles in the icosahedron

By fitting equilateral triangles into the edges, the surface of the icosahedron is formed (Fig. 6.13).

In Fig. 6.14, regular pentagons are fitted into the edges. There are twelve possibilities for this. The pentagons intersect each other. The resulting polyhedron is called a large dodecahedron. It is one of the four Kepler-Poinsot solids.

We can also fit pentagrams into the space diagonals (Fig. 6.15).

The polyhedron consisting of all twelve pentagrams is called a dodecahedron star. It is one of the four Kepler-Poinsot solids.

We have already seen that we can also fit Golden rectangles (Fig. 6.4). When dividing the Golden rectangles through their diagonals, four isosceles triangles are formed, two acute-angled and two obtuse-angled. Fig. 6.16 and Fig. 6.17 show the situation with the acute-angled and obtuse-angled isosceles triangles respectively.

Finally, we can also fit large equilateral triangles into the space diagonals (Fig. 6.18). There are twenty such triangles. The resulting star is called a large icosahedron. It is one of the four Kepler-Poinsot solids.

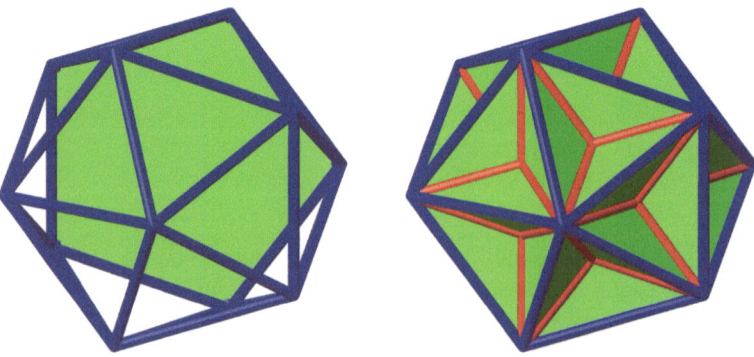

Fig. 6.14 Large dodecahedron

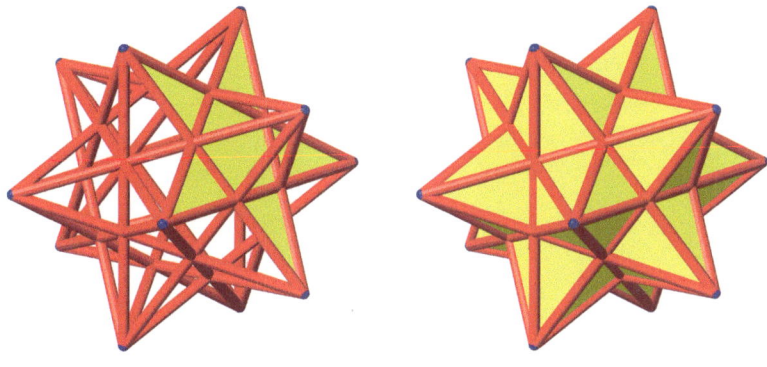

Fig. 6.15   Dodecahedron star

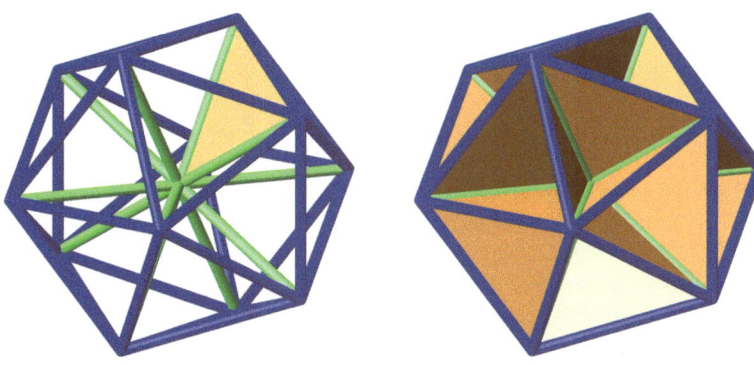

Fig. 6.16   Acute-angled isosceles triangles

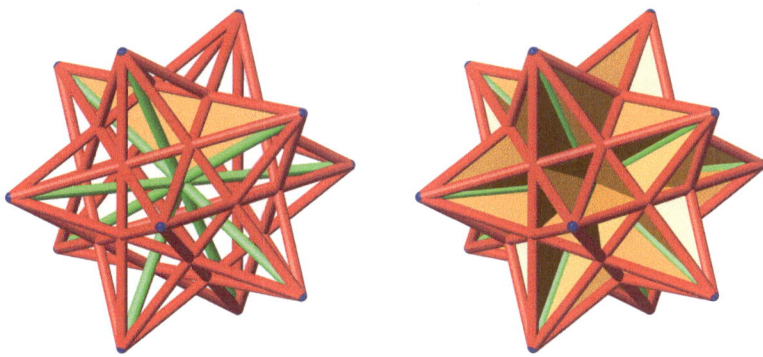

Fig. 6.17   Obtuse-angled isosceles triangles

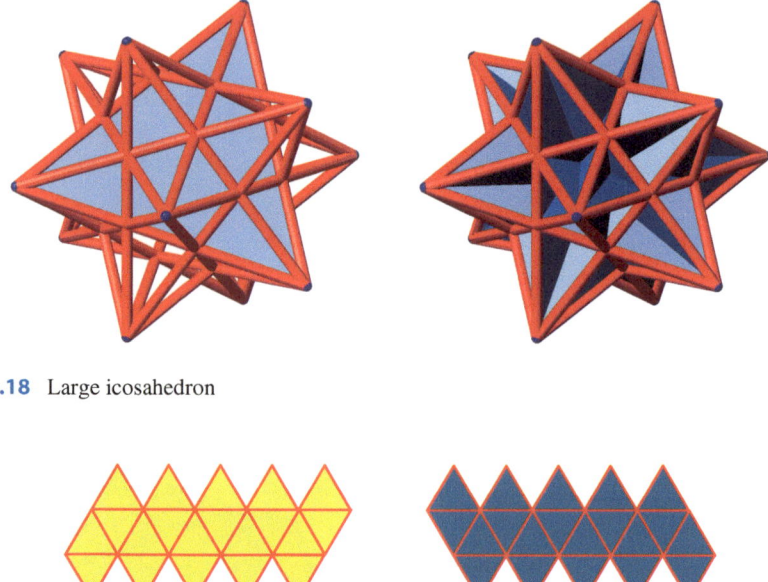

Fig. 6.18 Large icosahedron

Fig. 6.19 Unfolding of the Icosahedron

Another approach to the great icosahedron is achieved by unfolding the icosahedron (Fig. 6.19). The unfolded form (sometimes the unfolded form is also referred to as a net) of the icosahedron consists of twenty equilateral triangles. We assume that the unfolded form is yellow on one side and blue on the other.

Figure 6.20 illustrates the unfolding process up to the icosahedron. One of the triangles (the frontmost in the middle row) is held in place. At the common edges between two triangles, the same angle is rotated.

Fig. 6.20 Unfolding
(►https://doi.org/10.1007/000-bh8)

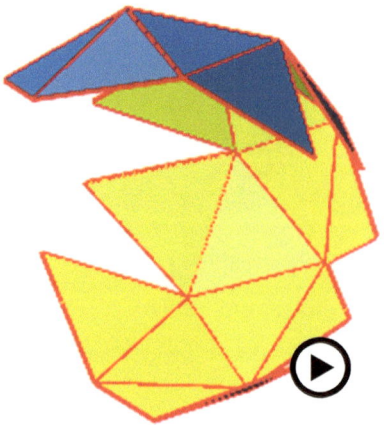

Fig. 6.21 Continuing to rotate
(▸https://doi.org/10.1007/000-bh5)

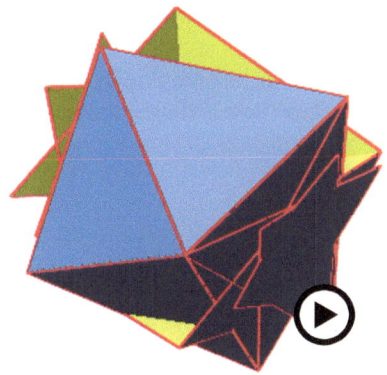

The idea now is to simply continue rotating upon reaching the icosahedron (Fig. 6.21). This results in natural intersections and overlaps. This is not possible with a real paper model.

After the icosahedron, an octahedron appears. Some side triangles of the icosahedron overlap. Then we see a tetrahedron and then, very small, a large icosahedron. Finally, everything folds into a blue triangle with a twenty-fold overlay like a folded yardstick. But it continues. The blue triangle seemingly changes color to yellow. In reality (that is, in our virtual reality), we are dealing with a surface penetration. Imagine in a book the sheet with the page numbers 329 and 330 and the following sheet with the page numbers 331 and 332. Now the two sheets penetrate each other, so that first the sheet with the page numbers 331 and 332 comes and only then the sheet with the page numbers 329 and 330. Fig. 6.22 illustrates this self-penetration of the equilateral triangles.

Now it goes back through the bottom in the self-unfolding of Fig. 6.21. The inside comes out. We first see a small-format large icosahedron, then a tetrahedron, then an octahedron, an icosahedron, and finally again the unfolded form in the starting position.

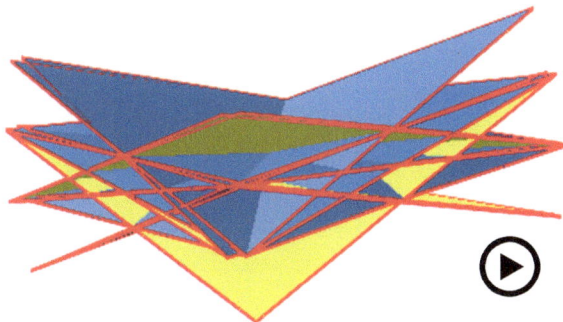

Fig. 6.22 Self-penetration (▸https://doi.org/10.1007/000-bh6)

6.3.2   Diagonals in a Dodecahedron

The center and the twenty vertices of a dodecahedron can be described in a Cartesian coordinate system with the following coordinates:

$$B_0 = (0, 0, 0) \tag{6.5}$$

$$
\begin{aligned}
B_1 &= \left(\Phi, 0, \tfrac{1}{\Phi}\right) & B_2 &= \left(\Phi, 0, -\tfrac{1}{\Phi}\right) & B_3 &= \left(-\Phi, 0, \tfrac{1}{\Phi}\right) & B_4 &= \left(-\Phi, 0, -\tfrac{1}{\Phi}\right) \\
B_5 &= \left(0, \tfrac{1}{\Phi}, \Phi\right) & B_6 &= \left(0, \tfrac{1}{\Phi}, -\Phi\right) & B_7 &= \left(0, -\tfrac{1}{\Phi}, \Phi\right) & B_8 &= \left(0, -\tfrac{1}{\Phi}, -\Phi\right) \\
B_9 &= \left(\tfrac{1}{\Phi}, \Phi, 0\right) & B_{10} &= \left(\tfrac{1}{\Phi}, -\Phi, 0\right) & B_{11} &= \left(-\tfrac{1}{\Phi}, \Phi, 0\right) & B_{12} &= \left(-\tfrac{1}{\Phi}, -\Phi, 0\right) \\
B_{13} &= (1, 1, 1) & B_{14} &= (1, 1, -1) & B_{15} &= (1, -1, 1) & B_{16} &= (1, -1, -1) \\
B_{17} &= (-1, 1, 1) & B_{18} &= (-1, 1, -1) & B_{19} &= (-1, -1, 1) & B_{20} &= (-1, -1, -1)
\end{aligned}
$$

Figure 6.23 shows the position of the points in the coordinate system with the vertex numbers. The dodecahedron is standing on the edge B_6B_8(unstable equilibrium).

Figure 6.24 shows the dodecahedron with all diagonals. Since we have twenty vertices, there are $\binom{20}{2} = 190$ connections between these vertices. However, the 30 edges of the dodecahedron are also counted.

Table 6.2 provides an overview of the edges and diagonals.

With the exception of the center diagonals, we can project the diagonals from the center onto the circumsphere (Fig. 6.25 in two special views).

Now we fill the edges of Fig. 6.24 with pentagons (Fig. 6.26) and obtain the dodecahedron in the usual form as a polyhedron. We have twelve pentagons.

But we can also span large pentagons (Fig. 6.27). There are again twelve pentagons, but no two pentagons share an edge.

Fig. 6.23 Vertex numbers in the dodecahedron

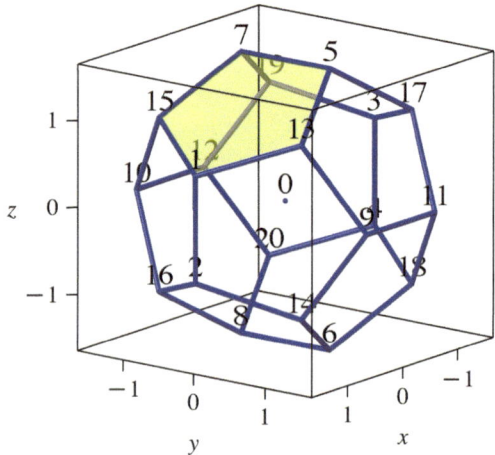

Fig. 6.24 Edges and diagonals

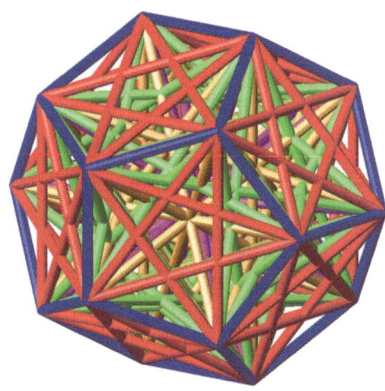

Tab. 6.2 Edges and diagonals in the dodecahedron

Farbe	Anzahl	Länge exakt	Länge numerisch	
	30	$\dfrac{2}{\Phi}$	1,236067978	Kante
	60	2	2	Seitenflächendiagonale
	60	$2\sqrt{2}$	2,828427124	Raumdiagonale
	30	2Φ	3,236067976	Raumdiagonale
	10	$2\sqrt{3}$	3,464101616	Mittelpunktdiagonale

Edges and Diagonals

In the face diagonals, we can inscribe squares and cubes (Fig. 6.28). There are 30 squares and five cubes. The ratio of the volume of a cube to the volume of the dodecahedron is $\frac{2}{5}(3 - \Phi) \approx 0{,}553$. The diagonal planes in the cube are rectangles with the aspect ratio $\sqrt{2} : 1$. They therefore have the DIN format.

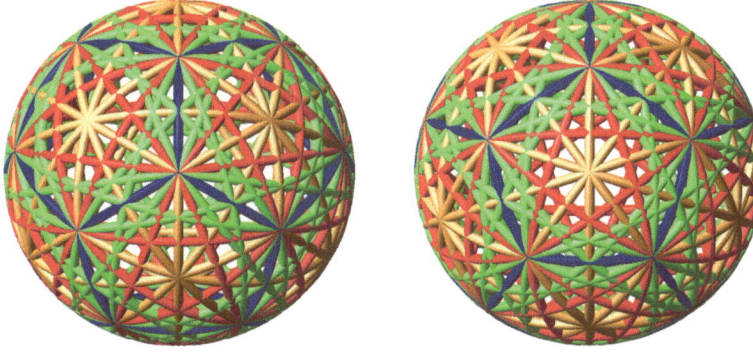

Fig. 6.25 Sphere projection

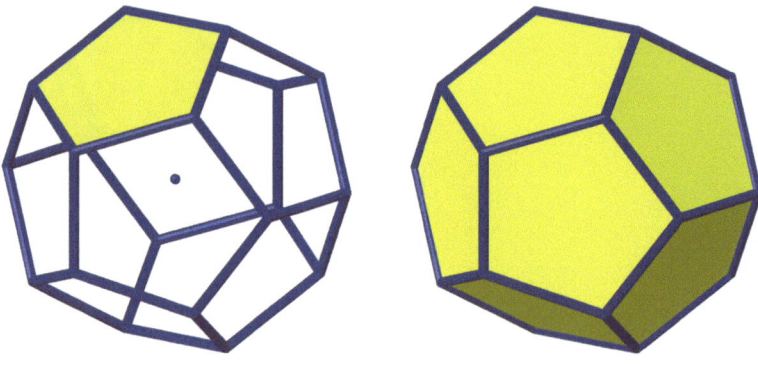

Fig. 6.26 Dodecahedron

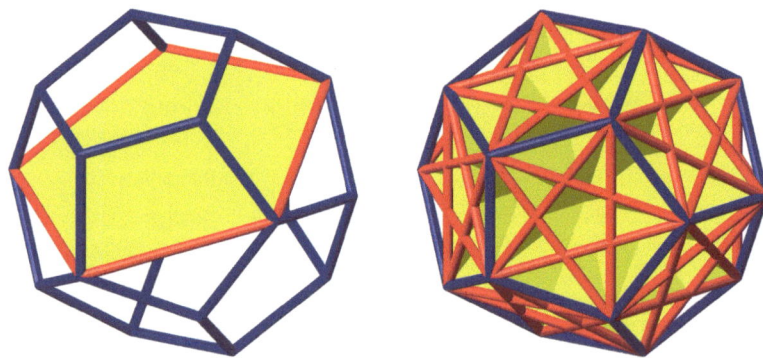

Fig. 6.27 Large Pentagons

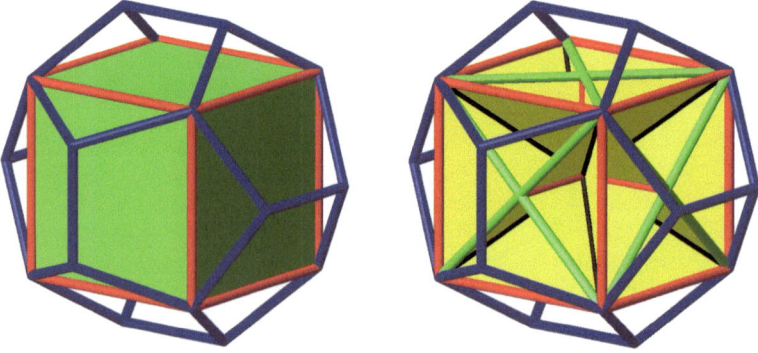

Fig. 6.28 Squares, cubes, and DIN format

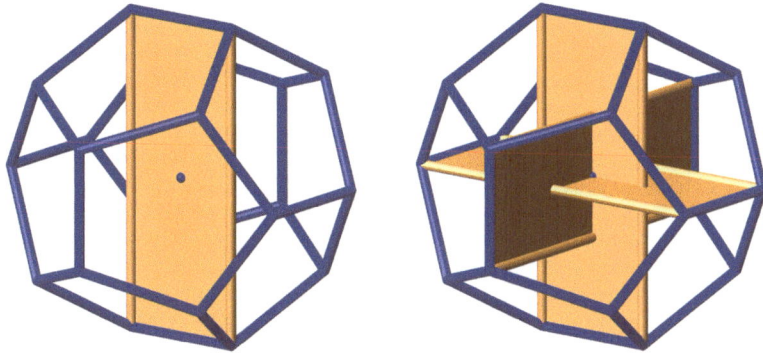

Fig. 6.29 High golden rectangles

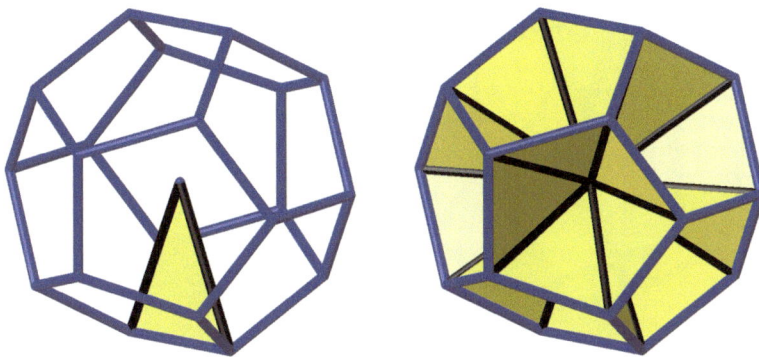

Fig. 6.30 Acute isosceles triangles

Finally, we also find rectangles with the aspect ratio $\Phi : \frac{1}{\Phi}$, thus high golden rectangles (Fig. 6.29).

We now divide the high golden rectangles with their diagonals. This results in four isosceles triangles, two acute and two obtuse. The tips of these isosceles triangles lie in the center of the dodecahedron (Fig. 6.30).

The figure can be built as a paper model (Fig. 6.31).

With the obtuse isosceles triangles, a star figure without volume is obtained (Fig. 6.32).

There are two ways to fit equilateral triangles into the diagonals. The smallest possible triangles have the face diagonals of the dodecahedron as sides (Fig. 6.33 and Fig. 6.34).

There are twenty such triangles, so one could speak of an icosahedron. However, the figure is not closed, so it is not a polyhedron. This becomes clear when rotating (Fig. 6.34).

With the largest possible equilateral triangles, tetrahedra and Kepler stars can be formed (Fig. 6.35).

Fig. 6.31 Paper model

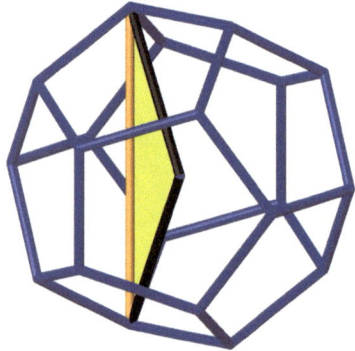

Fig. 6.32 Obtuse isosceles triangles

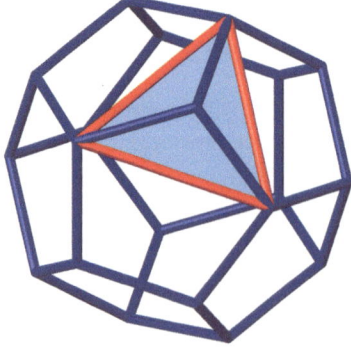

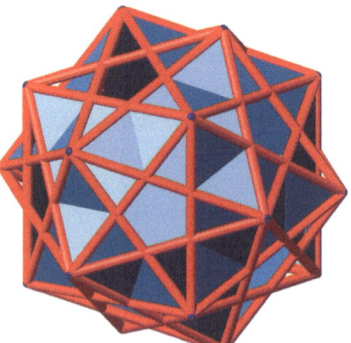

Fig. 6.33 Equilateral triangles in the dodecahedron

Fig. 6.34 Twenty
equilateral triangles
(▸https://doi.org/10.1007/000-bh7)

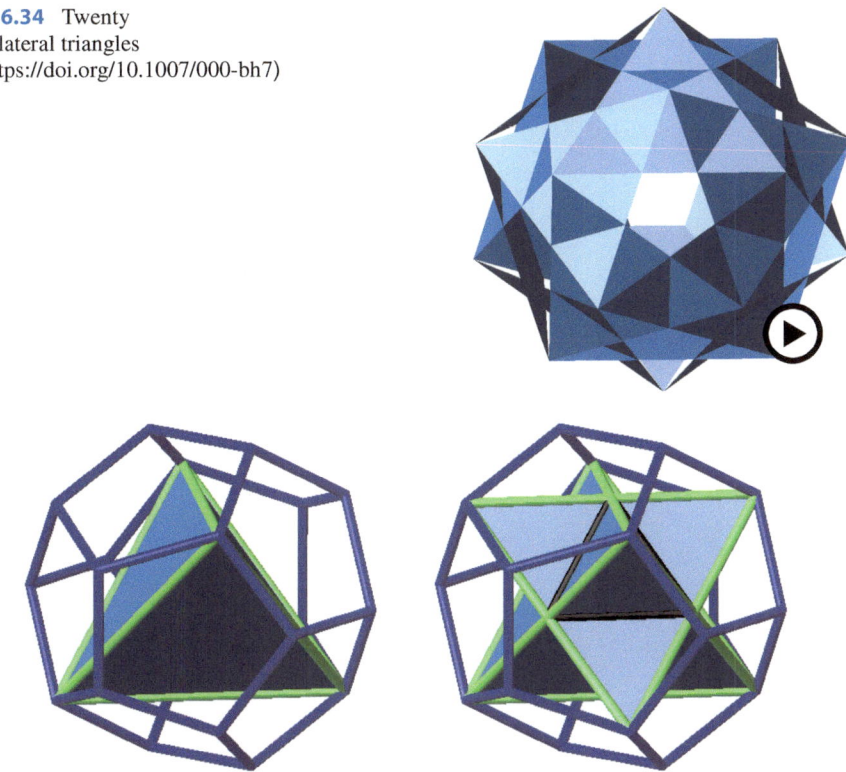

Fig. 6.35 Equilateral triangles, tetrahedra and Kepler star

There are a total of 40 such triangles (Fig. 6.36) and thus ten tetrahedra and five Kepler stars.

As already with the pentagons (Fig. 6.26 and Fig. 6.27) we can also fit two types of pentagrams into the diagonal framework of the dodecahedron. The small pentagrams lie on the faces of the dodecahedron (Fig. 6.37).

The large pentagrams (Fig. 6.38) form the icosahedron star. It is one of the four Kepler-Poinsot bodies.

6.4   Rhombic Solids

Under rhombic bodies we understand figures which are exclusively bounded by congruent rhombuses. The simplest example is the cube, which is even bounded by squares. With six equal rhombuses, a distorted cube can be built, a so-called rhombohedron.

Fig. 6.36 Forty
equilateral triangles
(▶https://doi.org/10.1007/000-bh4)

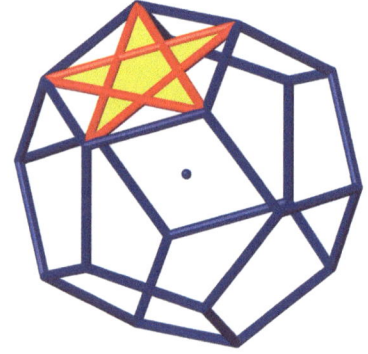

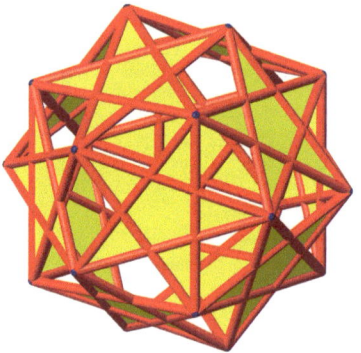

Fig. 6.37 Pentagrams

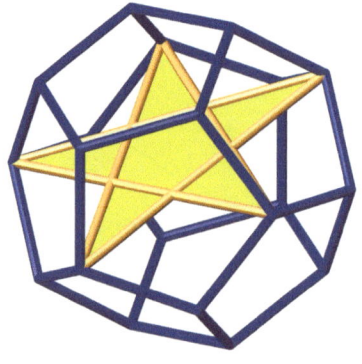

Fig. 6.38 Icosahedron star

6.4.1   The rhombic triacontahedron

By combining the twelve vertices of the icosahedron with the twenty vertices of the dodecahedron with the coordinates according to Eq. 6.4 and Eq. 6.5, we obtain 32 vertices of a rhombic body, the so-called rhombic triacontahedron (Fig. 6.39). It is bounded by 30 rhombi, hence the name. Triaconta is the Greek term for 30.

The diagonals of the side rhombi are in the ratio of the golden section, they can therefore be referred to as Golden Rhombi. The long diagonals (Majore) form an icosahedron, the short ones (Minore) a dodecahedron. For the acute angle α of the rhombi, the following results:

$$\alpha = \arctan(2) \approx 63{,}435° \tag{6.6}$$

The rhombic triacontahedron can be constructed as a braided model from six zig-zag strips (Fig. 6.40).

These zigzag strips are composed of ten Golden Rhombi. For technical reasons, two additional overlapping rhombi are added for stapling. Fig. 6.41 shows the layout of such a strip.

The rhombic triacontahedron can be decomposed into rhombic hexahedra with six Golden Rhombi as side faces. Interestingly, there are two different rhombic hexahedra with exactly the same six side rhombi, namely a "blunt" version (blue in Fig. 6.42) and a "sharp" version (red in Fig. 6.42).

In the blunt Golden Rhombic Hexahedron, there are two diametrical vertices (on the axes of rotation) where only blunt rhombic angles meet. In the sharp Golden Rhombic Hexahedron, there are correspondingly two diametrical vertices with only sharp rhombic angles. Fig. 6.43 shows the layouts (nets) of the two Golden Rhombic Hexahedra.

The two golden rhombohedra have the same surface, each consisting of six golden rhombi. However, the volumes are unequal, the volume of the acute golden rhombohedron is larger. The volumes are in the ratio of the golden section. Thus,

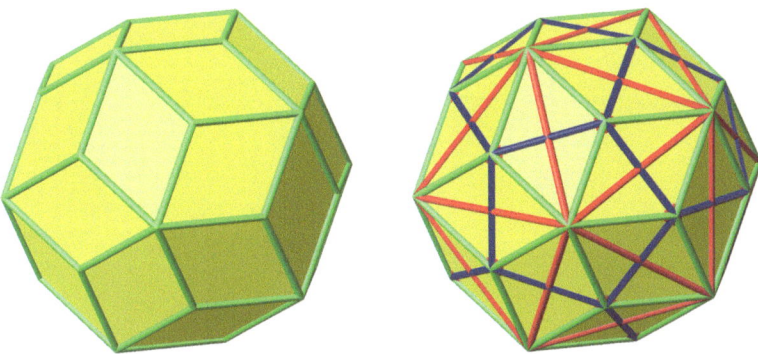

Fig. 6.39 Rhombic triacontahedron

Fig. 6.40 Braided model of
the rhombic triacontahedron

Fig. 6.41 Zigzag strip for the rhombic triacontahedron

Fig. 6.42 Rhombic hexahedra
(▶https://doi.org/10.1007/000-bh9)

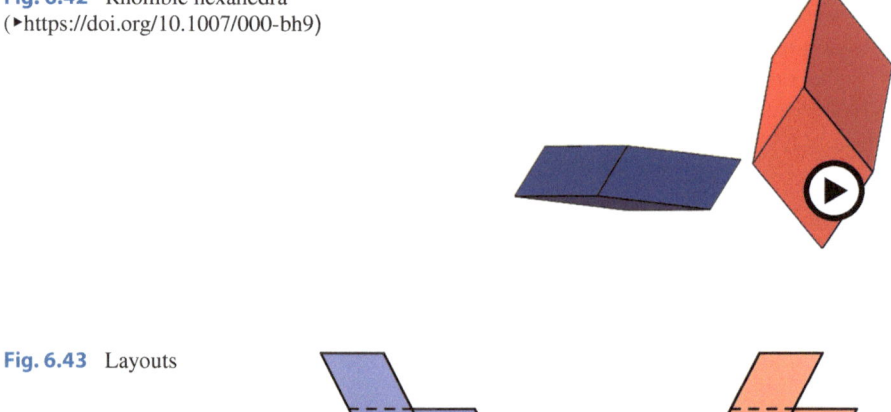

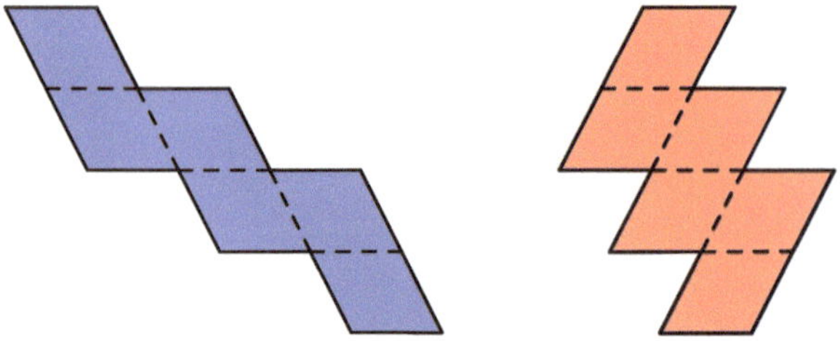

Fig. 6.43 Layouts

the acute golden rhombohedron is the major in terms of volume, and the obtuse golden rhombohedron is correspondingly the minor.

The rhombic triacontahedron can be decomposed into ten obtuse and ten acute golden rhombohedra each (Fig. 6.44) (cf. [18]).

6.4.2   Rhombic dodecahedron and rhombic icosahedron

In the dismantling and assembly process of the rhombic triacontahedron, there are two beautiful intermediate stages. After four steps, we obtain a solid bounded by twelve golden rhombi (Fig. 6.45 left). It consists of two obtuse and two acute golden rhombohedra. This body is referred to as the rhombic dodecahedron of the second type or also as Bilinski's rhombic dodecahedron (cf. [3]). The addition "second type" is necessary to distinguish it from the usual rhombic dodecahedron, whose rhombi have a diagonal ratio $\sqrt{2} : 1$ like the aspect ratio in DIN format. The rhombic dodecahedron of the second type was described by Bilinski in 1960. Bilinski showed that it is a space filler like the usual rhombic dodecahedron. The space can be filled completely and without overlap with rhombic dodecahedra of the second type.

After ten assembly steps with the installation of five obtuse and five acute golden rhombohedra each, the rhombic icosahedron results (Fig. 6.45 right). It has twenty rhombi on its surface.

6.4.3   A star solid

Into the "holes" of the icosahedron model of Fig. 6.16, 20 acute golden rhombohedra fit. Fig. 6.46 shows the associated movement process for the rhombohedra.

This creates a rhombic star with 20 tips (Fig. 6.47).

Fig. 6.44 Decomposition of the rhombic triacontahedron (▶https://doi.org/10.1007/000-bha)

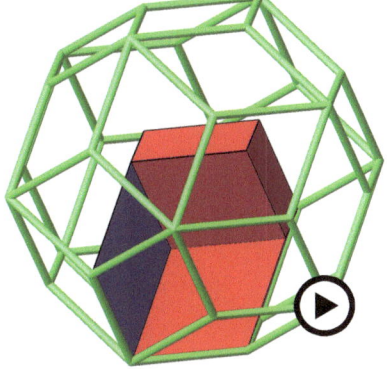

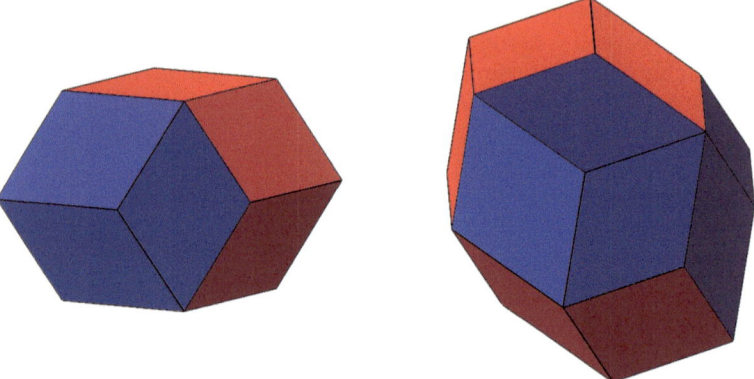

Fig. 6.45 Bilinski's rhombic dodecahedron and rhombic icosahedron

Fig. 6.46 Twenty acute
golden rhombohedra
(▸https://doi.org/10.1007/000-bhb)

Fig. 6.47 Rhombic star

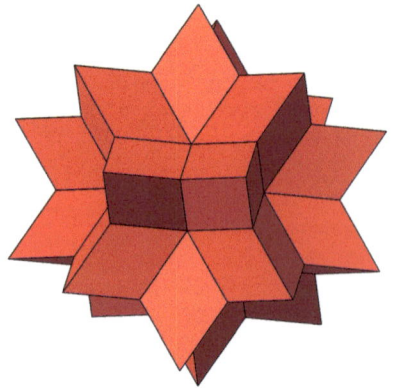

Examples

7

7.1   Number Games

Example 1: A beautiful formula:

$$e^{\operatorname{arcsinh}\left(\frac{1}{2}\right)} = \Phi \tag{7.1}$$

Example 2: Which positive numbers have the same decimal places as their reciprocal? We are looking for positive numbers that are smaller than their reciprocal by a natural number n, so:

$$x + n = \frac{1}{x}, n \in \mathbb{N} \tag{7.2}$$

As positive solutions of $x^2 + nx - 1 = 0$ we find the values:

$$x = \frac{-n + \sqrt{n^2 + 4}}{2} \tag{7.3}$$

For the reciprocals we find:

$$\frac{1}{x} = \frac{n + \sqrt{n^2 + 4}}{2} \tag{7.4}$$

The Table 7.1 shows some solutions, including the trivial case with $n = 0$ included. The Golden Ratio appears in the first non-trivial case.

Supplementary Information The online version contains supplementary material available at https://doi.org/10.1007/978-3-662-69890-7_7.

Table 7.1 Zahl und Kehrwert

n	Zahl	Kehrwert
0	1	1
1	$\frac{-1+\sqrt{5}}{2} = \frac{1}{\Phi} \approx 0,6180$	$\frac{1+\sqrt{5}}{2} = \Phi \approx 1,6180$
2	$-1 + \sqrt{2} \approx 0,4142$	$1 + \sqrt{2} \approx 2,4142$
3	$\frac{-3+\sqrt{13}}{2} \approx 0,3028$	$\frac{3+\sqrt{13}}{2} \approx 3,3028$
4	$-2 + \sqrt{5} \approx 0,2361$	$2 + \sqrt{5} \approx 4,2361$

Zahl und Kehrwert

Table 7.2 Number and square

n	Zahl	Quadratzahl
0	1	1
1	$\frac{1+\sqrt{5}}{2} = \Phi \approx 1,6180$	$\frac{3+\sqrt{5}}{2} = \Phi^2 \approx 2,6180$
2	2	4
3	$\frac{1+\sqrt{13}}{2} \approx 2,3028$	$\frac{7+\sqrt{13}}{2} \approx 5,3028$
4	$\frac{1+\sqrt{17}}{2} \approx 2,5616$	$\frac{9+\sqrt{17}}{2} \approx 6,5616$

Number and square number

Example 3: Which positive numbers have the same decimal places as their square (see [19], p. 173)? The condition $x^2 = x + n, n \in \mathbb{N}$, leads to

$$x = \frac{1 + \sqrt{1 + 4n}}{2} \tag{7.5}$$

and thus to the solutions of Table 7.2. Here too, the golden ratio is the first non-trivial example.

7.2   Circles and semicircles

Example 1: Semicircles in the Golden Rectangle (Fig. 7.1).

Examples 2 and 3: Square and three circles in the semicircle (Fig. 7.2).

Example 4: Semicircles and circle in the square (Fig. 7.3).

Example 5: To an equilateral Triangle we first draw the circumcircle and supplement with semicircles over the triangle sides to crescents. Then we find the Golden Section according to Fig. 7.4.

Example 6: We attach rectangles in the DIN format in the DIN format to the legs of a right-angled triangle, that is, with the aspect ratio $\sqrt{2} : 1$ (Fig. 7.5). We then

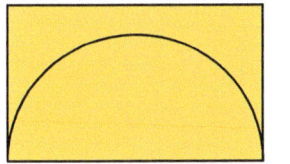 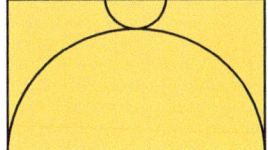 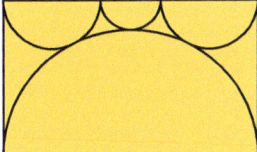

Fig. 7.1 Semicircles in the Golden Rectangle

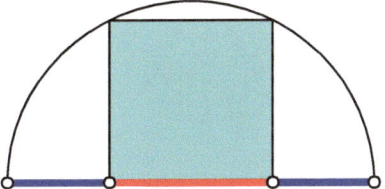 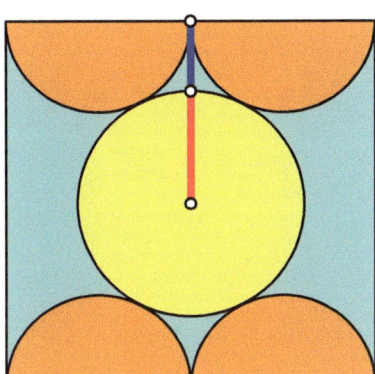

Fig. 7.2 Figures in the Semicircle

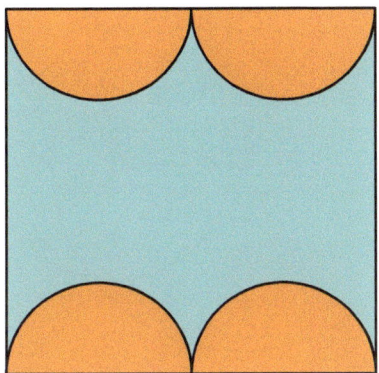

Fig. 7.3 In the Square

draw two circles. Their centers are the vertices with the acute angles of the right-angled triangle, and the radii are the diagonals of the respective attached DIN rectangle.

In the situation of Fig. 7.5, the two circles have two intersection points. We now vary the right-angled triangle (Fig. 7.6).

The two intersection points, as soon as they exist, lie on a circle that can be located with the golden ratio in the figure.

Figure 7.7 on the left shows the situation where the two circles touch on the left. The right-angled triangle, supplemented by a rectangle, is a high golden rectangle with the aspect ratio $\Phi^2 : 1$. The radii of the two circles also behave like $\Phi^2 : 1$.

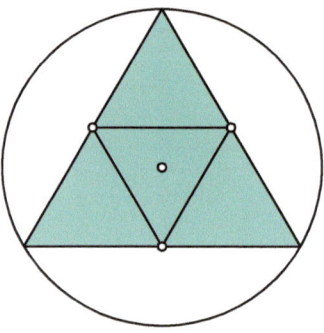
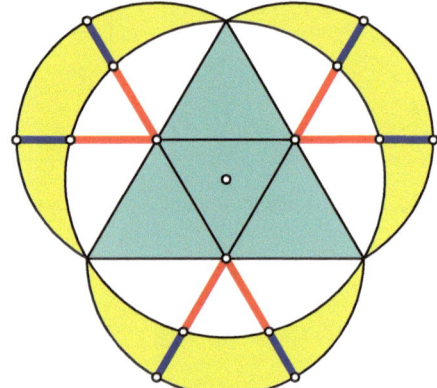

Fig. 7.4 Crescents

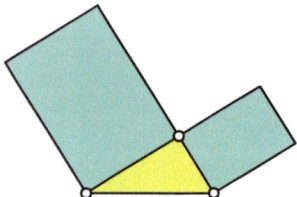
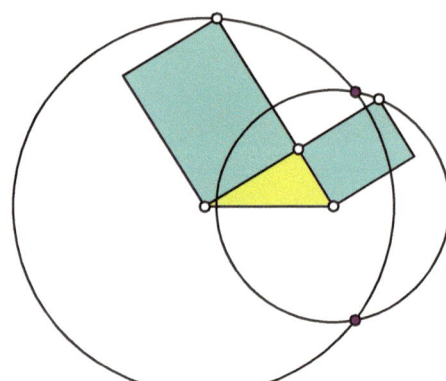

Fig. 7.5 Rectangles in DIN format

Fig. 7.6 Change of the triangle
(▶ https://doi.org/10.1007/000-cbj)

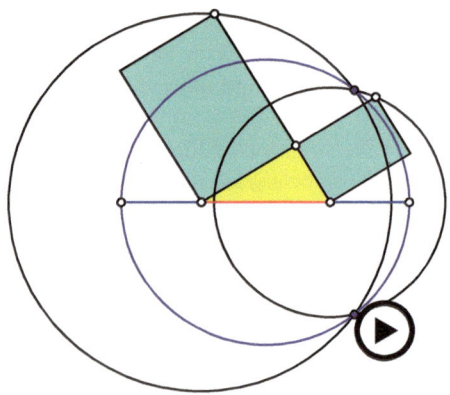

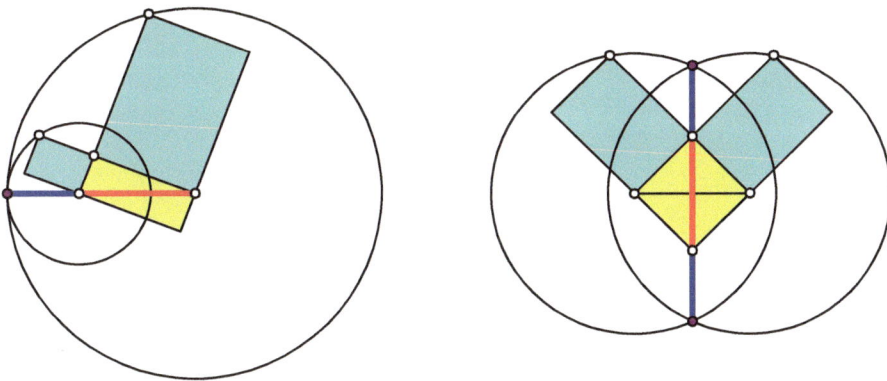

Fig. 7.7 Special cases

In the symmetrical case (Fig. 7.7 on the right), the triangle is right-angled isosceles and can be supplemented to a square.

Example 7: We start with five circles, whose radii from outside to inside form a geometric sequence with the quotient $\frac{1}{\Phi}$ (Fig. 7.8). The sequence can be constructed using pentagrams.

We can arrange these circles successively touching. The second and third circle fit as Major and Minor next to each other in the first circle (Fig. 7.9). By swapping the order, we get the opportunity to also fit the next circle. The fit of these circles results directly from a basic equation of the Golden Ratio:

$$1 = \frac{1}{\Phi} + \left(\frac{1}{\Phi}\right)^2 \tag{7.6}$$

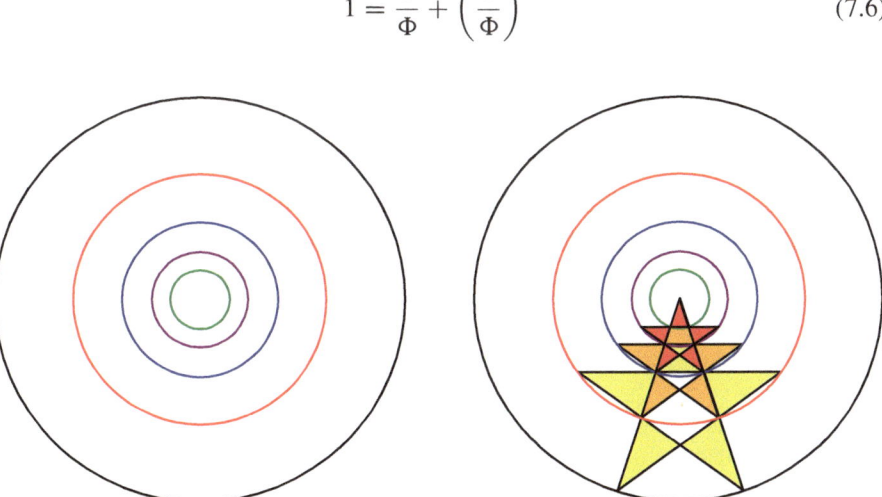

Fig. 7.8 Circle sequence. Construction pentagrams

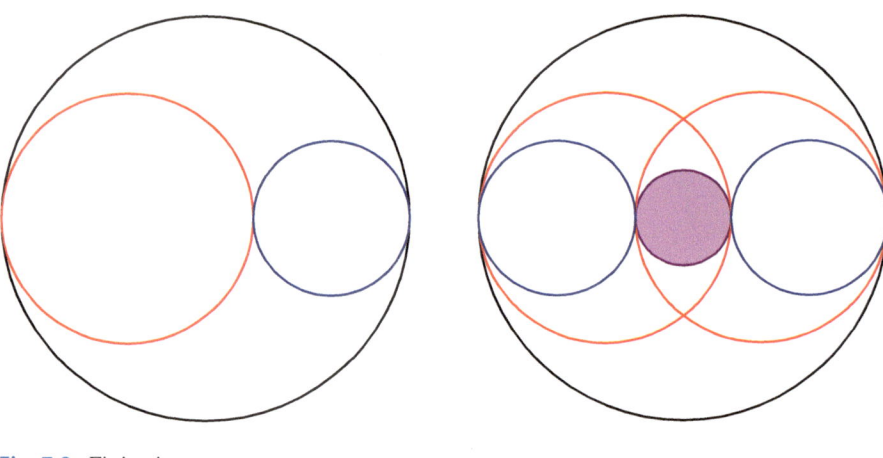

Fig. 7.9 Fitting in

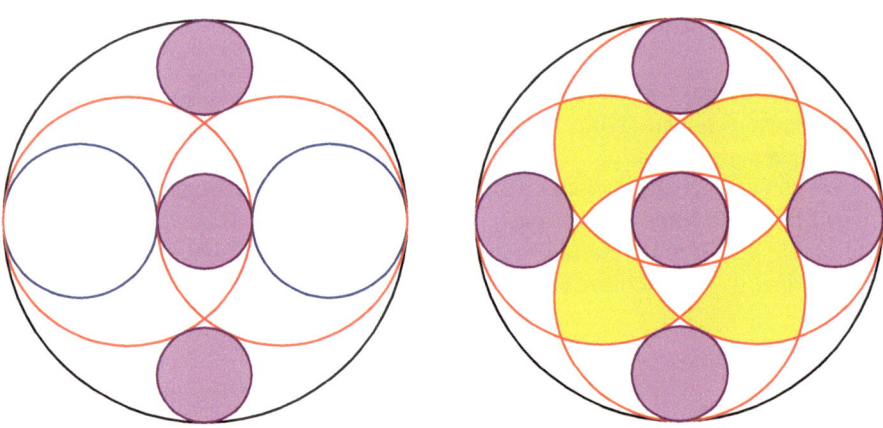

Fig. 7.10 Small circles

And now comes a first surprise: the small purple circle in the middle can also be fitted in touching at the top and bottom (Fig. 7.10). This is not trivial, but can be verified with calculations.

We now leave out the blue circles and rotate the red and purple circles around the center of the figure by 90°. Four spaces limited by red circles are created, which resemble Gothic tracery windows. In Fig. 7.10, these are marked in yellow.

And now comes the second surprise: we can fit four more purple circles into the tracery windows (Fig. 7.11). The nine purple circles form an exact square 3 × 3 grid. This is also not trivial.

And now we can fit in twelve more green circles. For the four green circles between the purple circles, the fit results from the properties of the Golden Ratio.

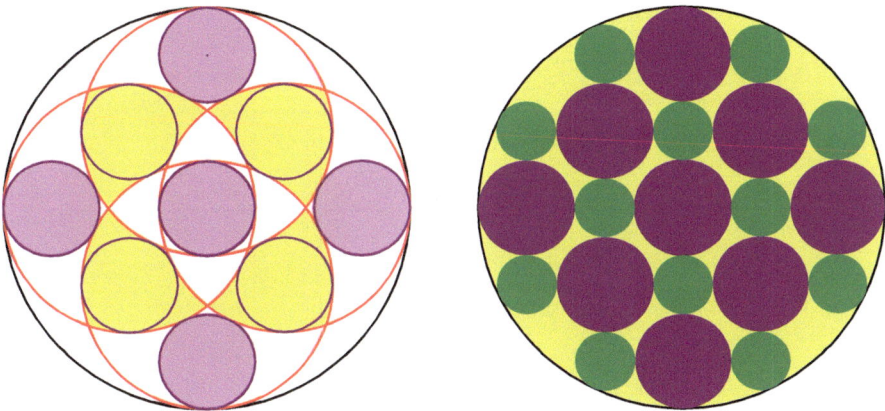

Fig. 7.11   Small circles in the grid

For the eight green circles, which also touch the black border circle, the fitting is again a surprise. The green circles are also arranged in a square grid.

7.3   Intersections

Example 1: The parabola with the equation $y = x^2 - 1$ intersects the line with the equation $y = x$ (Fig. 7.12 left) at the points with the coordinates (Φ, Φ) and $\left(-\frac{1}{\Phi}, -\frac{1}{\Phi}\right)$. The origin of coordinates divides the distance between the two intersection points in the ratio $\Phi^2 : 1$ (Golden Section).

Example 2: The hyperbola with the equation $y = 1 + \frac{1}{x}$ intersects the line with the equation $y = x$ (Fig. 7.12 right) at the points with the coordinates (Φ, Φ) and $\left(-\frac{1}{\Phi}, -\frac{1}{\Phi}\right)$.

Example 3: The hyperbola with the equation $y = 1 + \frac{1}{x}$ intersects the parabola with the equation $y = x^2 - 1$ at the points with the coordinates (Φ, Φ), $\left(-\frac{1}{\Phi}, -\frac{1}{\Phi}\right)$ and $(-1, 0)$.

Example 4: The intersection points of graphs of different Chebyshev polynomials often have coordinates with the Golden Ratio. Thus, the graphs of $T_1(x) = x$ and $T_4(x) = 8x^4 - 8x^2 + 1$ have four intersection points with the coordinates $(1, 1)$, $\left(-\frac{1}{2}, -\frac{1}{2}\right)$, $\left(\frac{1}{2\Phi}, \frac{1}{2\Phi}\right)$, $\left(-\frac{\Phi}{2}, -\frac{\Phi}{2}\right)$. The graphs of $T_2(x) = 2x^2 - 1$ and $T_3(x) = 4x^3 - 3x$ have three intersection points with the coordinates $(1, 1)$, $\left(\frac{1}{2\Phi}, -\frac{\Phi}{2}\right)$, $\left(-\frac{\Phi}{2}, \frac{1}{2\Phi}\right)$ (Fig. 7.13).

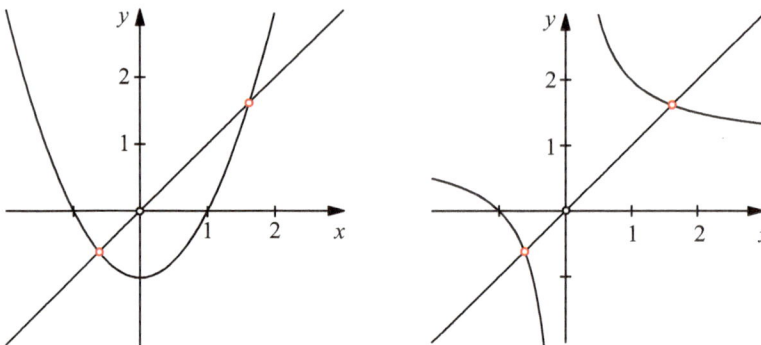

Fig. 7.12 Parabola and Hyperbola

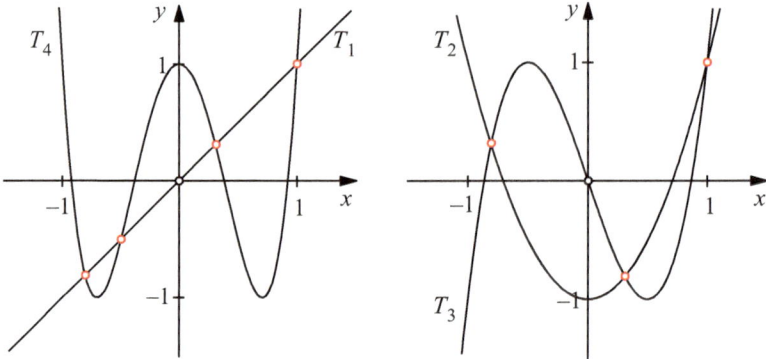

Fig. 7.13 Chebyshev Polynomials

Example 5: The circle $x^2 + y^2 = 3$ and the hyperbola $xy = 1$ intersect at the four points with the coordinates $\left(\Phi, \frac{1}{\Phi}\right)$, $\left(\frac{1}{\Phi}, \Phi\right)$, $\left(-\Phi, -\frac{1}{\Phi}\right)$, $\left(-\frac{1}{\Phi}, -\Phi\right)$ (Fig. 7.14 left).

Example 6: The two parabolas in the square (Fig. 7.14 right) have the four intersection points with the coordinates $(2, 2)$, $\left(-\Phi, \frac{1}{\Phi}\right)$, $(-1, -1)$, $\left(\frac{1}{\Phi}, -\Phi\right)$. The four intersection points lie on a circle.

7.4   Area and Volume

Example 1: A square is divided according to Fig. 7.15 so that the three outer triangles are equal in area.

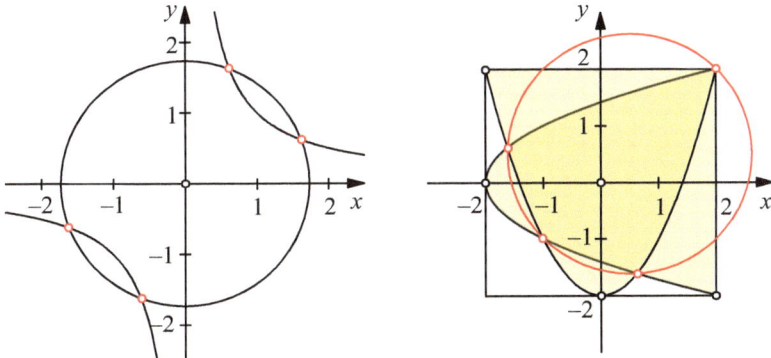

Fig. 7.14 Conic sections and circles

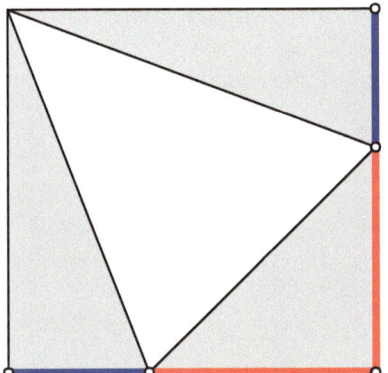

Fig. 7.15 Equal area outer triangles

Example 2: The area between the parabolic arc $y = x^2, x \in [0, 1]$, the x axis and the straight line $x = 1$ is to be divided by two vertical cutting lines at a distance of $\frac{1}{2}$ such that the section between the two cutting lines makes up half the area (Fig. 7.16).

The entire area has the content:

$$\int_0^1 x^2 dx = \frac{1}{3} \qquad (7.7)$$

The two vertical cutting lines have the equations $x = x_0$ and $x = x_0 + \frac{1}{2}$. The area condition yields:

$$\frac{1}{3}\left(x_0 + \frac{1}{2}\right)^3 - \frac{1}{3}x_0^3 = \frac{1}{6} \qquad (7.8)$$

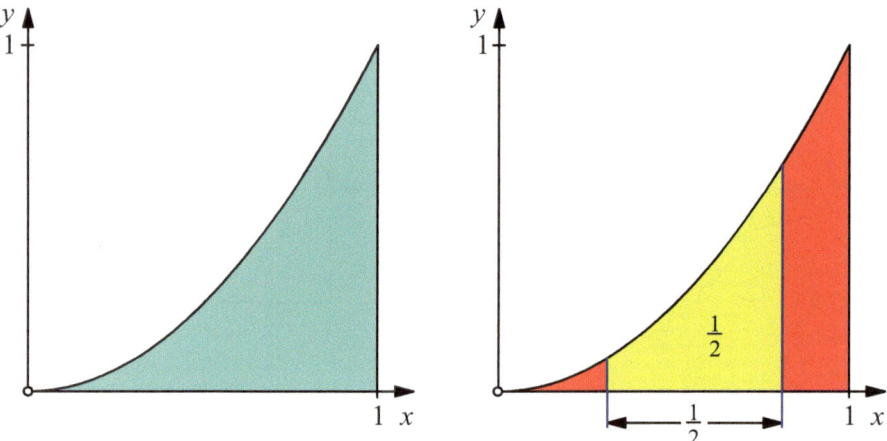

Fig. 7.16 Area halving

In this equation, the cubic part drops out and a quadratic equation remains:

$$4x_0^2 + 2x_0 - 1 = 0 \tag{7.9}$$

This has the following positive solution:

$$x_0 = \frac{-1 + \sqrt{5}}{4} = \frac{1}{2}\frac{1}{\Phi} \approx 0{,}3090 \tag{7.10}$$

For the second cutting line, we obtain:

$$x_0 + \frac{1}{2} = \frac{1 + \sqrt{5}}{4} = \frac{1}{2}\Phi \approx 0{,}8090 \tag{7.11}$$

Example 3: A cross formed from five squares is cut by a square with an area equal to the cross (Fig. 7.17).

The equality of area can be illustrated by a common decomposition (Fig. 7.18).

Example 4: A pyramid with a rectangular base is divided by a plane passing through a base edge into two equal-volume parts (Fig. 7.19).

Example 5: The area within the Archimedean spiral $r(\varphi) = a\varphi$, $\varphi \in [0, 2\pi]$, is to be halved by a line originating from the center (Fig. 7.20) (Task from [10]).

For the area element, the following applies:

$$dS = \frac{1}{2}a^2\varphi^2 d\varphi \tag{7.12}$$

Fig. 7.17 Cross and Square

Fig. 7.18 Common Decomposition
(▶ https://doi.org/10.1007/000-bhd)

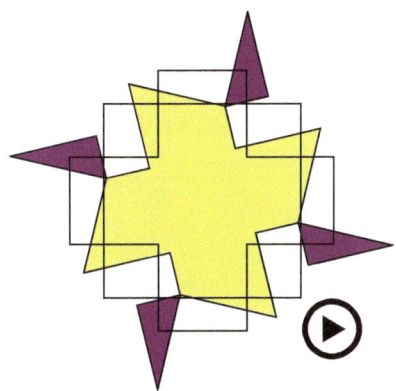

Therefore, the full spiral area is:

$$S_1 = \frac{1}{2}a^2 \int_0^{2\pi} \varphi^2 \, d\varphi = \frac{1}{6}a^2 (2\pi)^3 \qquad (7.13)$$

We denote the polar angle of the dividing line as φ_0. From this, the half spiral area results:

$$S_{\frac{1}{2}} = \frac{1}{2}a^2 \int_{\varphi_0}^{\varphi_0 + \pi} \varphi^2 \, d\varphi = \frac{1}{6}a^2 (\varphi_0 + \pi)^3 - \frac{1}{6}a^2 \varphi_0^3 \qquad (7.14)$$

Thus, we have the condition $(\varphi_0 + \pi)^3 - \varphi_0^3 = \frac{1}{2}(2\pi)^3$. From this, the following results:

$$\varphi_0 = \pi \frac{-1 + \sqrt{5}}{2} = \frac{\pi}{\Phi} \qquad (7.15)$$

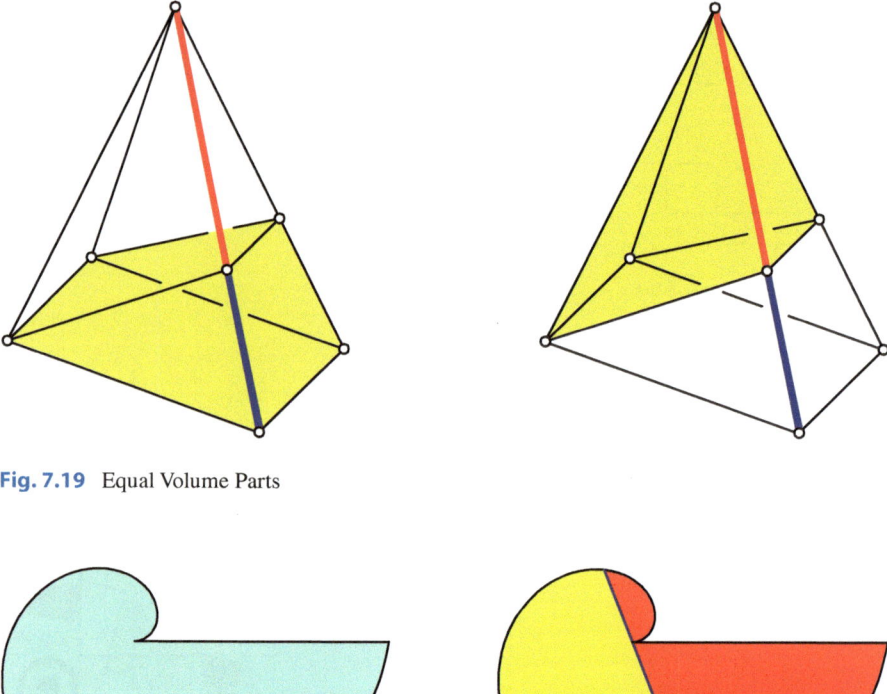

Fig. 7.19 Equal Volume Parts

Fig. 7.20 Halving of the spiral area

Example 6: We shorten the height of a cylinder by $p\%$ and increase the radius by $p\%$. With the notation $x = \frac{p}{100}$ the changed volume results from the existing cylinder volume $V_0 = \pi r^2 h$:

$$V_1 = \pi(r(1 + x))^2(h(1 - x)) \tag{7.16}$$

So, we have the following change factor to discuss:

$$f(x) = (1 + x)^2(1 - x), x \in [0, 1] \tag{7.17}$$

Figure 7.21 shows the corresponding function graph.

We obtain the maximum volume increase for $x = \frac{1}{3}$, so for $p \approx 33,33\%$. The question now is whether there is a percentage p that does not lead to a change in volume. To do this, we work on the following equation:

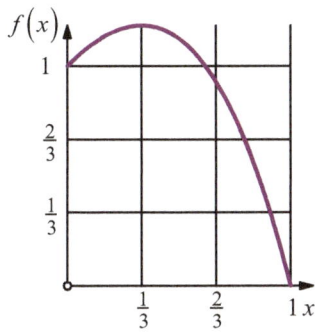
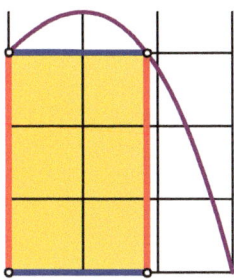

Fig. 7.21 Function graph

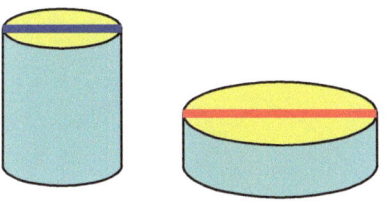
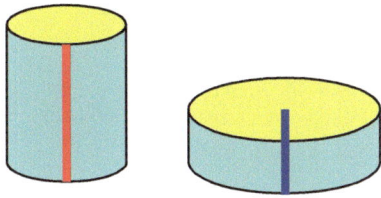

Fig. 7.22 Volume-equivalent cylinders

$$(1 + x)^2(1 - x) = 1 \tag{7.18}$$

It has the solutions $\left\{0, \frac{1}{\Phi}, -\Phi\right\}$. The solution zero is trivial, those who do nothing, do nothing wrong. The positive solution in the permissible range is $x = \frac{1}{\Phi}$, so $p \approx 61,8\%$.. Accordingly, we can fit a Golden Rectangle into the function graph (Fig. 7.21).

Figure 7.22 shows an arbitrary original cylinder and a changed, but volume-equivalent Cylinder.

The diameter of the changed cylinder is the major compared to the diameter of the original cylinder. However, the height of the changed cylinder is not the minor compared to the height of the original cylinder, but the minor of the minor.

7.5   Constructions with Equilateral Triangles

Example 1: The Fig. 7.23 and the Fig. 7.24 show a construction of the Golden Ratio and the Golden Rectangle based on equilateral Triangles.

Example 2: A Reuleaux Triangle is created from an equilateral triangle by attaching circular arcs, whose centers are the triangle corners. It has the same diameter everywhere, so like the circle it is a so-called "constant width". We now divide

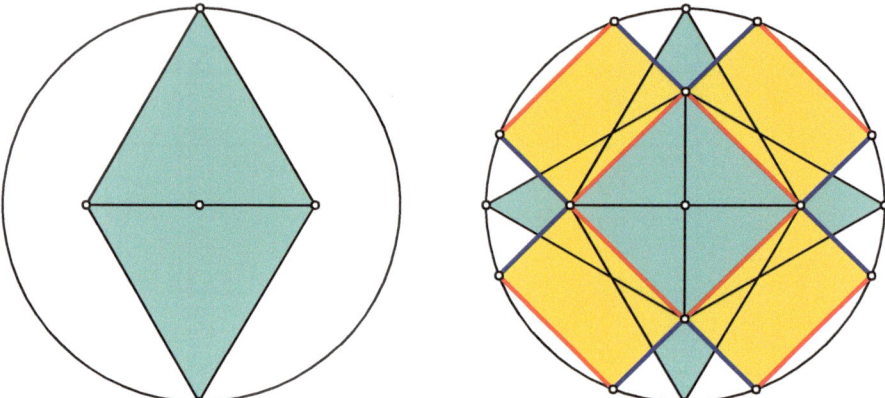

Fig. 7.23 Equilateral Triangles

Fig. 7.24 Step by Step
(▸ https://doi.org/10.1007/000-bhc)

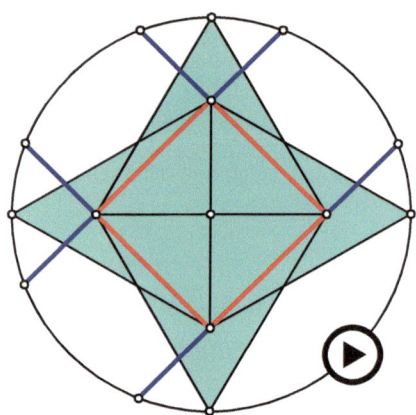

the equilateral triangle in a Reuleaux triangle into four equal parts (Fig. 7.25) and rotate a partial triangle to the stop. This results in a division in the Golden Ratio.

7.6   Optimization

Example 1: In an isosceles triangle with a given fixed incircle, the base height can be changed (Fig. 7.26). In which situation is the leg length minimal? ([29], p. 299)

In the optimal situation, the Golden Ratio appears (Fig. 7.27).

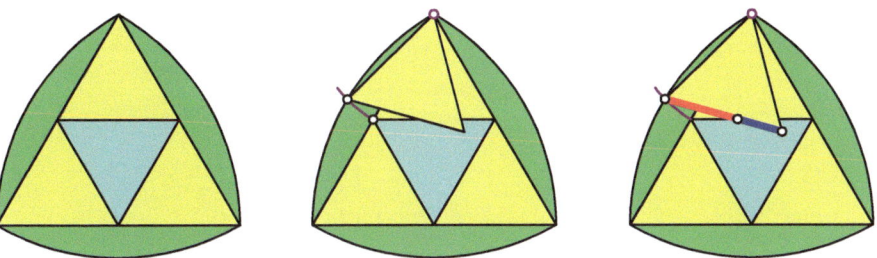

Fig. 7.25 Reuleaux Triangle and Golden Ratio

Fig. 7.26 Minimal leg length?
(▶ https://doi.org/10.1007/000-bhf)

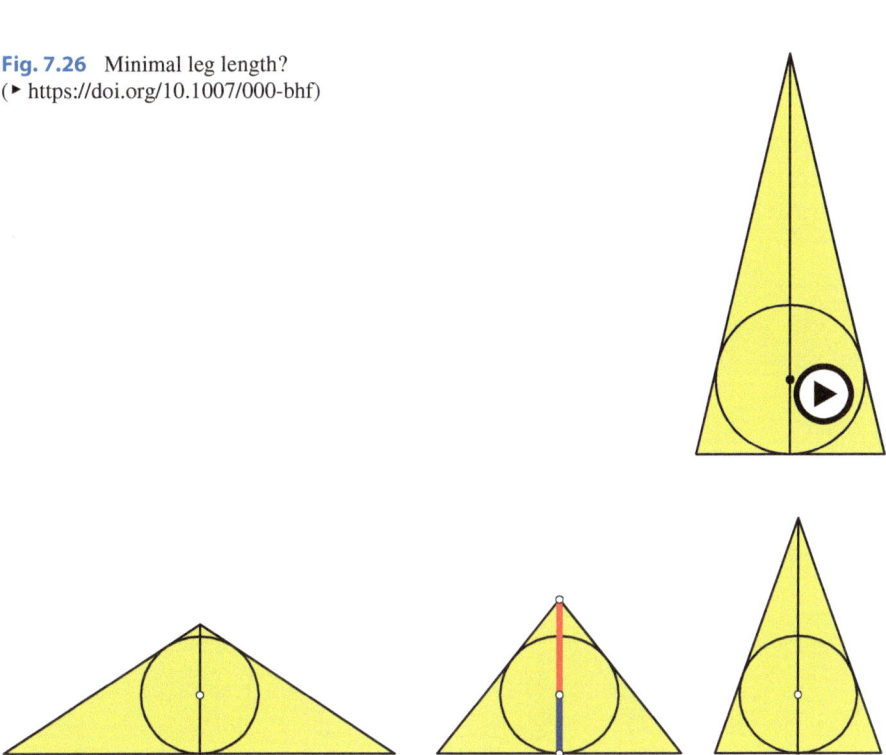

Fig. 7.27 Optimal Solution with Golden Ratio

Example 2: A rectangle is inscribed in a circle and then rotated around the center by a right angle (Fig. 7.28). In which situation is the area of the union of the two rectangles (cross area) maximum?

Figure 7.29 shows the cross area as a function of the given angle α, where the circle radius is set to one. The square (for $\alpha = \frac{\pi}{4}$) has the area $\frac{1}{2}$. However, it is not the solution of our optimization problem. For $\alpha = \arctan\left(\frac{1}{\Phi}\right)$ and $\alpha = \arctan(\Phi)$ a larger circle area results, namely $\frac{1}{\Phi}$.

In the optimal solution, we can fit a Golden Rectangle (Fig. 7.30).

Fig. 7.28 Cross in the circle
(▶ https://doi.org/10.1007/000-bhg)

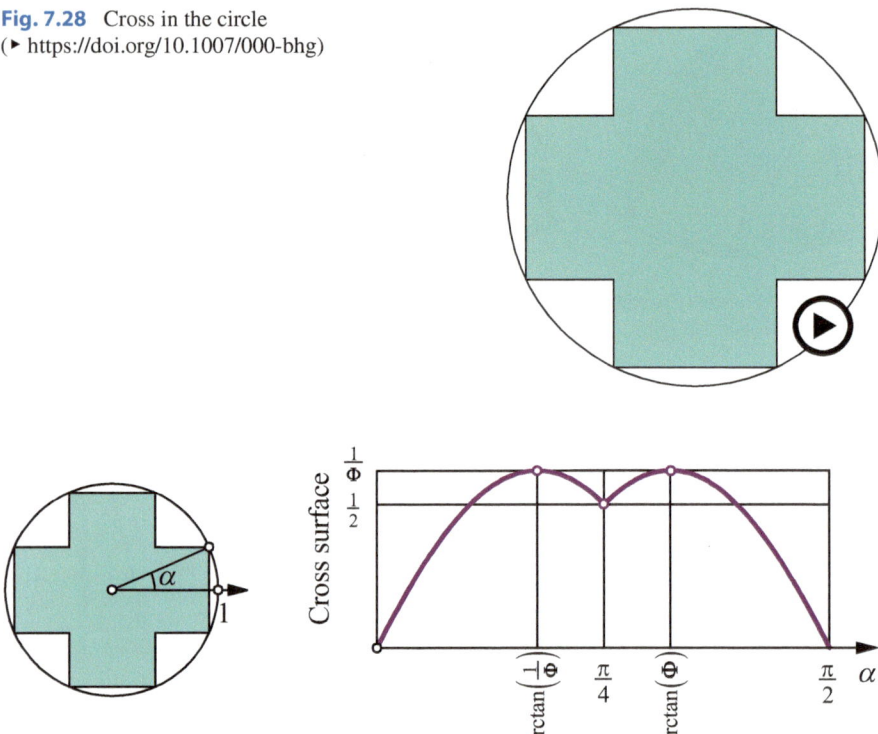

Fig. 7.29 Function graph circle area

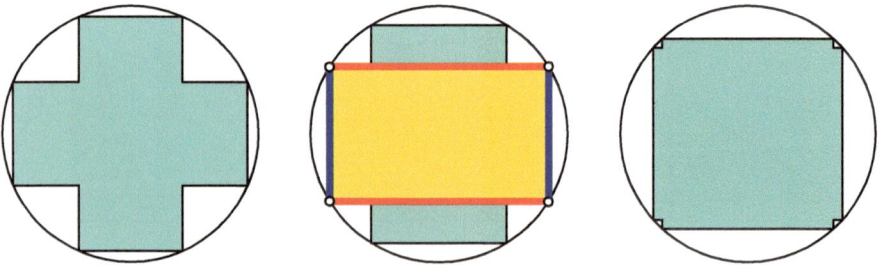

Fig. 7.30 Optimal solution in the Golden Ratio

7.7   Golden Probabilities

Example 1: First, an unfair game: Two people A and B alternately toss a coin, A starts. Whoever tosses "heads" first wins the game. The game is obviously unfair, as A has the greater chance of winning. Already in the first move, A serves half of the chances, leaving only a quarter of the chances for B in the corresponding first

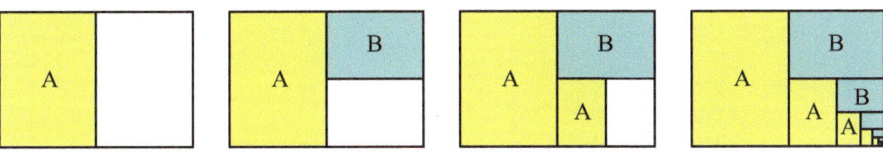

Fig. 7.31 Two to one chance distribution

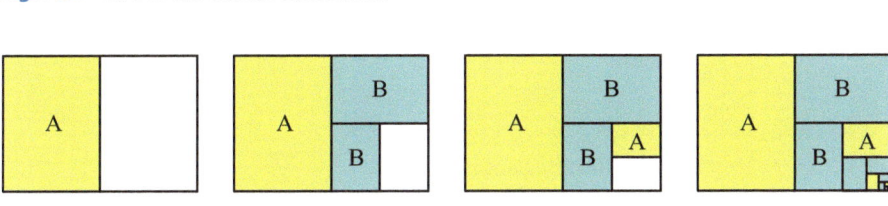

Fig. 7.32 Four to three chance distribution

move. Fig. 7.31 illustrates the chances of winning on a DIN-A4 paper, which can be continuously halved in terms of area.

The chance of winning for A is twice as large as that for B. A has a total chance of winning of two thirds.

We now change the game by still having A start, but B can throw twice in a row. That means, we use the game sequence ABBABBABB... (Fig. 7.32).

This is still unfair, and to the disadvantage of B. A has a total chance of winning of four sevenths.

Even with the game sequence ABBBABBBABBB... the game is unfair (Fig. 7.33).

The only fair game sequence would be that A starts with a coin toss, but then only B is allowed to toss the coin. Then, however, one could limit oneself to A's coin toss.

We return to the sequence ABBABBABB..., but replace the coin toss with the spinning of a wheel of fortune (Fig. 7.34). The wheel of fortune consists of two complementary sectors with the shares p and q. Of course, $p + q = 1$. The shares p and q correspond to the probabilities of landing in the green winning area or the complementary losing area when spinning the wheel of fortune.

In $p = \frac{1}{2}$ we have the same situation as with coin flipping. In our context, therefore, $p < \frac{1}{2}$ must be the case.

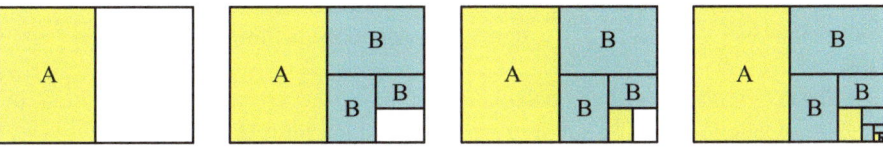

Fig. 7.33 Eight to seven chance distribution

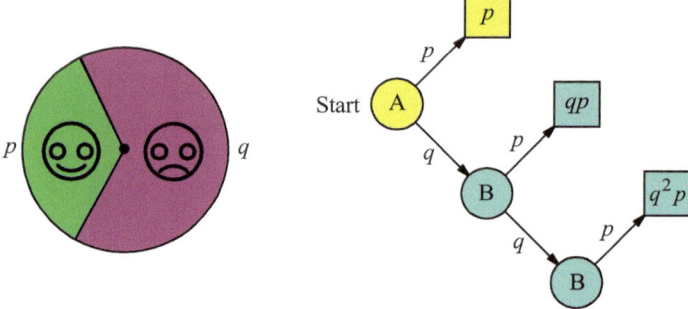

Fig. 7.34 Wheel of fortune and first game period

The game sequence ABBABBABB… is periodic with the period ABB. If neither A nor B are successful in the first period, the situation is the same as at the start. For a fair game, this means that the chances of winning must already be balanced in the first period. These chances of winning can be calculated as follows. If A wins right at the beginning, the game is over. This happens with a probability p. With the counter-probability q A is unsuccessful, and B gets to spin the wheel of fortune. If B wins, the game is also over. This happens, counted from the start, with the probability qp. If B fails on the first attempt, which happens with the probability q, B gets to spin a second time and then wins, counted from the start, with the probability q^2p, namely: failure of A, failure of B on the first attempt, success of B on the second attempt. Of course, B can also be unsuccessful on the second attempt (with a probability q), and then it's A's turn again and the second game period begins.

For the first game period, A has a chance of winning p and B has a total chance of winning $qp + q^2p$. For a fair game, this results in the condition:

$$p = qp + q^2p \tag{7.19}$$

Because of $p + q = 1$ the solution relevant to us results:

$$p = \left(\frac{1}{\Phi}\right)^2 \approx 0,3820, q = \frac{1}{\Phi} \approx 0,6180 \tag{7.20}$$

In a fair game, the sectors of the wheel of fortune have the ratio of the golden section (cf. [12]).

Thus, A has the winning chance $\left(\frac{1}{\Phi}\right)^2$ on the first attempt, then B has the winning chances $\left(\frac{1}{\Phi}\right)^3$ and $\left(\frac{1}{\Phi}\right)^4$, then again A has the winning chance $\left(\frac{1}{\Phi}\right)^5$ and so on.

These winning chances, calculated from the start, can be represented in a square of side length 1 by subdivision in the golden ratio (Fig. 7.35).

Fig. 7.35 Fair game

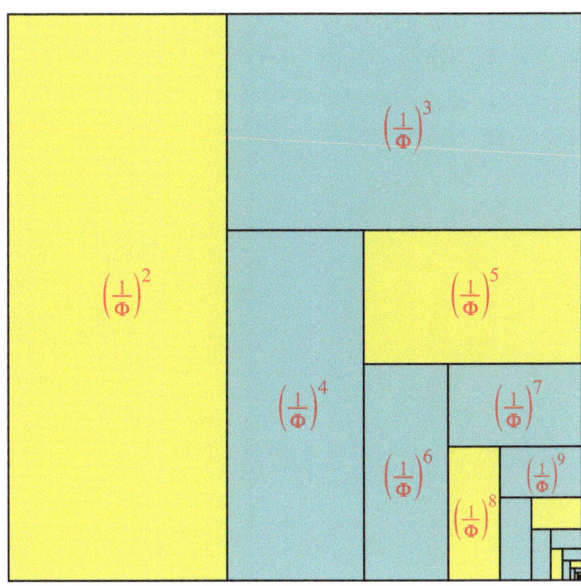

The subdivision consists of tall golden rectangles in portrait format and golden rectangles in landscape format. The areas of the individual rectangles are reduced step by step with the factor $\frac{1}{\Phi} \approx 0,6180$. Because of the fairness of the game, the yellow area of the square makes up exactly half of the square area. The author admits that he initially did not want to believe this based on appearances. However, we can arrange the yellow rectangles (now all in portrait format) into a spiral that exactly fills half of the square (Fig. 7.36).

From Fig. 7.36 we read off the relationship:

$$\left(\frac{1}{\Phi}\right)^2 + \left(\frac{1}{\Phi}\right)^5 + \left(\frac{1}{\Phi}\right)^8 + \left(\frac{1}{\Phi}\right)^{11} + \cdots = \frac{1}{2} \qquad (7.21)$$

Example 2: Unfortunately, it is not possible to make a game with the sequence ABABAB… fair by changing p. In the first game period AB, A would have the winning probability p and B would have the winning probability qp. The fairness condition $qp = p$ would result in $q = 1$ and $p = 0$. Thus, the game is uninteresting. However, we can change the game by choosing the winning sectors on the wheel of fortune for A and B to be complementary (Fig. 7.37).

In this case, A and B each have their own winning sector. The winning sector of A is smaller than that of B, which is intended to compensate for the fact that A gets to start. From the first game period AB, the fairness condition $p = q^2$ results, so:

$$p = (1 - p)^2 \qquad (7.22)$$

The solution relevant to us is $p = \left(\frac{1}{\Phi}\right)^2 \approx 0,3820$ and $q = \frac{1}{\Phi} \approx 0,6180$.

Fig. 7.36 Half area share

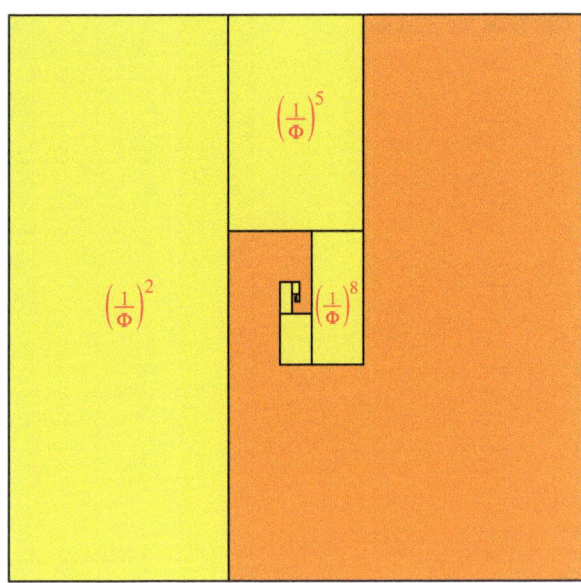

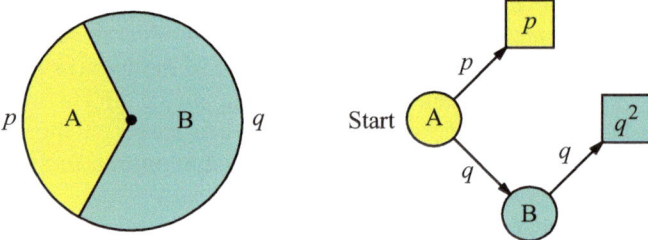

Fig. 7.37 Complementary winning sectors

Therefore, the wheel of fortune must be divided in the ratio of the golden section.

Figure 7.38 shows the further course of the game.

The individual chances of winning can be represented in the square with side length 1 (Fig. 7.39). The fairness of the game is reflected in the point symmetry of the figure.

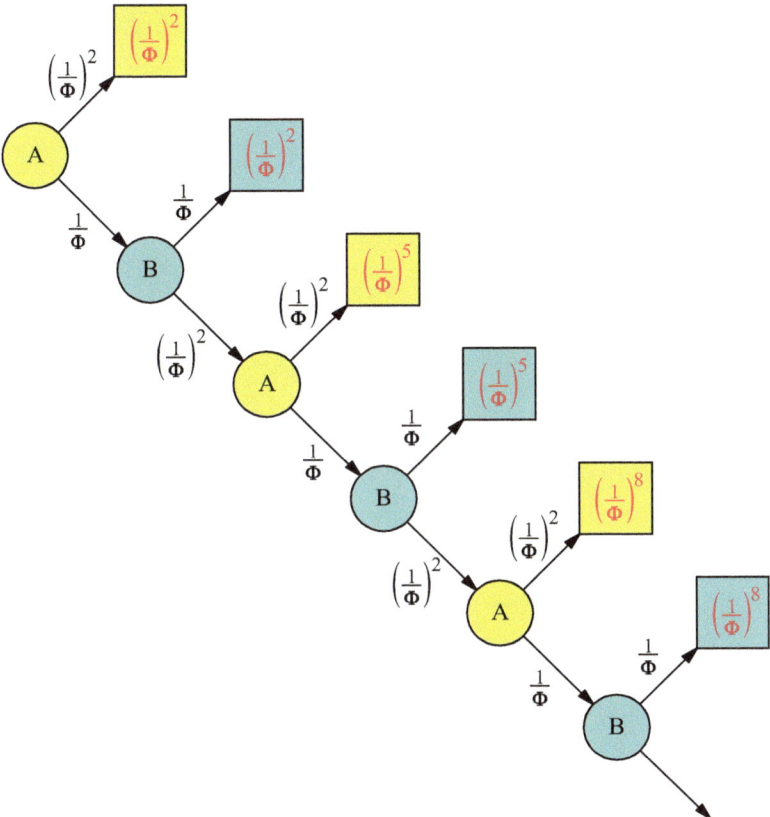

Fig. 7.38 Course of the game

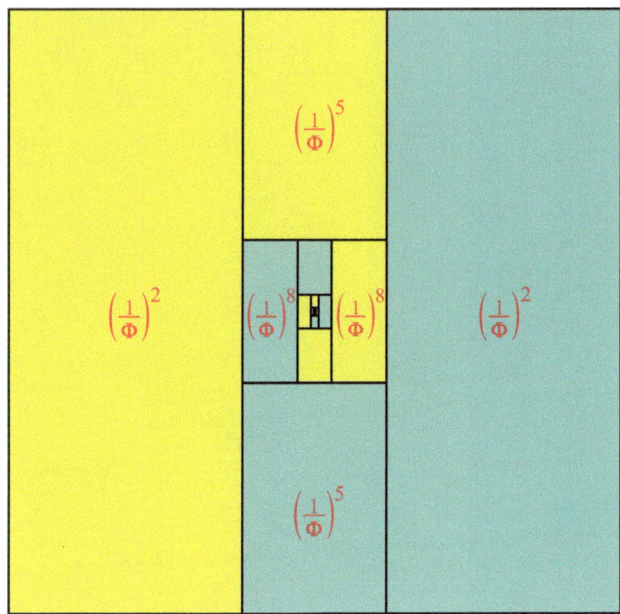

Fig. 7.39 Chances of winning

Literatur

1. Baptist, P. (Hrsg.). (2008). *Alles ist Zahl* (S. 2008). Universitäts-Verlag: Motive von Eugen Jost. Köln.
2. Beutelspacher, A. & Petri, B. (1996). Der Goldene Schnitt. 2. Aufl. Spektrum Akad. Verlag.
3. Bilinski, S. (1960). Über Rhombenisoeder. *Glasnik mat.-fiz. i astr., 15*(4), 251–262.
4. Coxeter, H. S. M. (1973). *Regular Polytopes. Third Edition.* Dover.
5. Coxeter, H. S. M. (1981). *Unvergängliche Geometrie. Wissenschaft und Kultur, Band 17.* 2. Aufl. Birkhäuser.
6. Doczi, G. (1996). *Die Kraft der Grenzen.* Engel.
7. Filler, A. (2012). Die Fibonacci-Folge von der Grundschule bis zum Abitur. *MU Der Mathematik-Unterricht, 1(58)*, 49–55.
8. Ghyka, M. C. (1959). *Le Nombre d'Or.* Gallimard.
9. Hagenmaier, O. (1989). *Der Goldene Schnitt. Ein Harmoniegesetz und seine Anwendung.* Weltbild Verlag. ISBN: 3926187417.
10. Heitzer, J. (1998). *Spiralen, ein Kapitel phänomenaler Mathematik.* Klett.
11. Hemenway, P. (2008). *Der geheime Code. Die rätselhafte Formel, die Kunst, Natur und Wissenschaft bestimmt.* Evergreen-Verlag.
12. Henze, N., & Schilling, J. (2019). Ein faires Glücksrad mit unterschiedlich großen Sektoren. *MU, Mathematikunterricht, 6*(2019), 33–39.
13. Herz-Fischler, R. (1998). *A mathematical history of the golden number.* Dover Publications.
14. Hilton, P., & Pedersen, J. (1994). *Build your own polyhedra.* Addison-Wesley.
15. Hoehn, A., & Huber, M. (2005). *Pythagoras. Erinnern Sie sich?* Orell Füssli-Verlag.
16. Hofstetter, K. (2002). A simple construction of the golden section. *Forum Geometricorum, 2,,* 65–66.
17. Huntley, H. E. (1970). *The divine proportion.* Dover.
18. Kowalewski, G. (1938). *Der Keplersche Körper und andere Bauspiele.* Leipzig: K.F. Koehlers Antiquarium.
19. Laugwitz, D. (1975). Die Quadratwurzel aus 5, die natürlichen Zahlen und der Goldene Schnitt. *Jahrbuch Überblicke Mathematik, 1975,* 173–181.
20. Lehmann, I. (2012a). Goldener Schnitt, Fibonacci-Zahlen und Goldene Figuren. In: Die Fibonacci-Zahlen und der goldene Schnitt. *MU Der Mathematik-Unterricht, 1*(58), 5–12.
21. Lehmann, I. (2012b). Goldener Schnitt, Fibonacci-Zahlen und Goldene Figuren. In: Die Fibonacci-Zahlen und der goldene Schnitt. *MU Der Mathematik-Unterricht, 1*(58), 39–48.
22. Müller-Sommer, H. (2012). Entdeckungen an der Goldenen Spirale. In: Die Fibonacci-Zahlen und der goldene Schnitt. *MU Der Mathematik-Unterricht, 1*(58), 24–27.
23. Müller, C. (2013). 50 Jahre Spezi in Jena. Ein mathematischer Blick auf eine ganz SPEZIelle Schule. BoD – Books on Demand GmbH. ISBN 978-3-7322-2973-4.
24. Neukirchner, T. (2012). Geometrie der Kettenbrüche. In: Die Fibonacci-Zahlen und der goldene Schnitt. *MU Der Mathematik-Unterricht, 1*(58), 13–18.

H. Walser, *The Golden Ratio*, https://doi.org/10.1007/978-3-662-69890-7

25. Neumann, O. (2004). Der Goldene Schnitt und seine Iteration. Einiges über Kunst und Zahlentheorie. Mathematik im Fluss der Zeit. Wolfgang Hein, Peter Ullrich (Hrsg.). *Algorismus, 44,* 355–366.

26. Pedoe, D. (2011). *Geometry and the visual arts.* Dover Publications.

27. Posamentier, A. S., & Lehmann, I. (2007). *The (fabulous) fibonacci numbers.* Prometheus Books.

28. Reis, H. (1990). *Der Goldene Schnitt.* Verlag für systematische Musikwissenschaft.

29. Reuter, D. (1984). „Goldene Terme" nicht nur am regulären Fünf- und Zehneck. *Praxis der Mathematik, 26,* 298–302.

30. Schmitz, M. (2013). Regelmäßige Drei- und Sechsecke aus Papierstreifen. Die Wurzel. *Zeitschrift für Mathematik, 6,* 130–135.

31. Schmitz, M. (2021). *MATHE KANNSTE KNICKEN. Kreativer und aktivierender Mathematikunterricht mit Papierfalten.* Seiten: 242. eISBN: 978-3-446-47153-5. Print ISBN: 978-3-446-46940-2. © 2021 Hanser.

32. Steibl, H. (1996). *Geometrie aus dem Zettelkasten.* Franzbecker. ISBN 3-88120-269-2.

33. Timerding, H. E. (1937). *Der Goldene Schnitt* (4. Aufl.). Teubner-Verlag.

34. Tropfke, J. (1980). *Geschichte der Elementarmathematik. Bd. 1, Arithmetik und Algebra.* 4. Aufl. de Gruyter.

35. Vincent, R. (1999). *Géométrie du nombre d'or.* 2e Aufl. Chalagam Edition.

36. Walser, H. (2011a). *Geometrische Miniaturen. Figuren – Muster – Symmetrien.* EAGLE. Edition am Gutenbergplatz.

37. Walser, H. (2011). Proof without words: Fibonacci trapezoids. *Mathematics Magazine, 84,* 295.

38. Walser, H. (2012a). *99 Schnittpunkte. Beispiele – Bilder – Beweise.* 2. Auflage. EAGLE 010. Leipzig: Edition am Gutenbergplatz.

39. Walser, H. (2012b). Fibonacci-Trapeze. In: Die Fibonacci-Zahlen und der goldene Schnitt. *MU Der Mathematik-Unterricht, 1*(58), 19–23.

40. Walser, H. (2015a). Vieleck-Knoten. MNU. Der mathematische und naturwissenschaftliche Unterricht. 68/4 (15. 7. 2015), S. 224–227. ISSN 0025-5866.

41. Walser, H. (2015b). Vielecke aus Streifen. *Der Falter // Magazin.* Origami Deutschland e. V., 64, Oktober 2015, S. 9–12.

42. Walser, H. (2021). Spiralen in Rechtecken. MI, Mathematikinformation Nr. 75, 15. September 2021. ISSN 1612-9156. 3–15.

43. Walser, H. (2022). *Spiralen, Schraubenlinien und spiralartige Figuren. Mathematische Spielereien in zwei und drei Dimensionen.* Springer.